Sensors for Performance Analysis in Team Sports

Sensors for Performance Analysis in Team Sports

Guest Editors

Bruno Emanuel Nogueira Figueira
Diogo Alexandre Martins Coutinho
Bruno Gonçalves

MDPI Basel • Beijing • Wuhan • Barcelona • Belgrade • Novi Sad • Cluj • Manchester

Guest Editors
Bruno Emanuel Nogueira Figueira
Departamento de Desporto e Saúde
Escola de Saúde e Desenvolvimento Humano
Universidade de Évora
Évora
Portugal

Diogo Alexandre Martins Coutinho
Department of Sports Sciences, Exercise and Health
University of Trás-os-Montes and Alto Douro
Maia
Portugal

Bruno Gonçalves
Departamento de Desporto e Saúde
Escola de Saúde e Desenvolvimento Humano
Universidade de Évora
Évora
Portugal

Editorial Office
MDPI AG
Grosspeteranlage 5
4052 Basel, Switzerland

This is a reprint of the Special Issue, published open access by the journal *Sensors* (ISSN 1424-8220), freely accessible at: https://www.mdpi.com/journal/sensors/special_issues/C9R20EEY34.

For citation purposes, cite each article independently as indicated on the article page online and as indicated below:

Lastname, A.A.; Lastname, B.B. Article Title. *Journal Name* **Year**, *Volume Number*, Page Range.

ISBN 978-3-7258-3589-8 (Hbk)
ISBN 978-3-7258-3590-4 (PDF)
https://doi.org/10.3390/books978-3-7258-3590-4

Contents

sensors

MDPI

Article

Validation of Aerobic Capacity (VO2max) and Pulse Oximetry in Wearable Technology

Bryson Carrier, Sofia Marten Chaves and James W. Navalta *

School of Integrated Health Sciences, Department of Kinesiology and Nutrition Sciences, University of Nevada, Las Vegas, NV 89154, USA

* Correspondence: james.navalta@unlv.edu

Abstract: Introduction: As wearable technology becomes increasingly popular and sophisticated, independent validation is needed to determine its accuracy and potential applications. Therefore, the purpose of this study was to evaluate the accuracy (validity) of VO2max estimates and blood oxygen saturation measured via pulse oximetry using the Garmin fēnix 6 with a general population participant pool. Methods: We recruited apparently healthy individuals (both active and sedentary) for VO2max (n = 19) and pulse oximetry testing (n = 22). VO2max was assessed through a graded exercise test and an outdoor run, comparing results from the Garmin fēnix 6 to a criterion measurement obtained from a metabolic system. Pulse oximetry involved comparing fēnix 6 readings under normoxic and hypoxic conditions against a medical-grade pulse oximeter. Data analysis included descriptive statistics, error analysis, correlation analysis, equivalence testing, and bias assessment, with the validation criteria set at a concordance correlation coefficient (CCC) > 0.7 and a mean absolute percentage error (MAPE) < 10%. Results: The Garmin fēnix 6 provided accurate VO2max estimates, closely aligning with the 15 s and 30 s averaged laboratory data (MAPE for 30 s avg = 7.05%; Lin's concordance correlation coefficient for 30 s avg = 0.73). However, it failed to accurately measure blood oxygen saturation (BOS) under any condition or combined analysis (MAPE for combined conditions BOS = 4.29%; Lin's concordance correlation coefficient for combined conditions BOS = 0.10). Conclusion: While the Garmin fēnix 6 shows promise for estimating the VO2max, reflecting its utility for both individuals and researchers, it falls short in accurately measuring BOS, limiting its application for monitoring acclimatization and managing pulmonary diseases. This research underscores the importance of validating wearable technology to leverage its full potential in enhancing personal health and advancing public health research.

Keywords: cardiorespiratory fitness; fitness tracker; activity monitor; biometric technology; altitude; hypoxia

Academic Editor: Javier Prieto

Received: 14 November 2024

Revised: 12 December 2024

Accepted: 22 December 2024

Published: 6 January 2025

Citation: Carrier, B.; Marten Chaves, S.; Navalta, J.W. Validation of Aerobic Capacity (VO2max) and Pulse Oximetry in Wearable Technology. *Sensors* **2025**, *25*, 275. https://doi.org/10.3390/s25010275

1. Introduction

Wearable technology (WT) has continued to grow in popularity and sophistication each year, with WT reaching the number one spot in worldwide surveys of fitness trends in seven of the last nine years and being in the top three for the other two years (2018 and 2021) [1–9]. According to recent surveys, almost one in three Americans use a wearable device to track their health and exercise, and around 70% of people own at least one wearable or plan to buy one in the next year [10,11]. This prevalence of WT may represent a revolutionary change in physiology and public health research simply due to the vast pool of potential data that may become available to researchers. Also, an important aspect is the constant monitoring of physiological metrics that these devices perform, which will provide granular details into a person's physiology that could transform human physiology research [12,13]. However, this transformation may only come to be realized if WT devices are found to be accurate in their measurements and estimates. As these consumer-grade wearable devices are not subject to any type of regulation, there is no governing body ensuring their accuracy. Thus, if researchers, athletes/coaches, public health officials, and healthcare professionals hope to continue to utilize these devices, an understanding of their accuracy and when they can appropriately be used is necessary. This underpins the importance of independent validation of WT devices by researchers to further several scientific fields.

Among the many variables that WT can estimate or measure, the maximal aerobic capacity (or VO2max) and blood oxygen saturation (BOS) measured via pulse oximetry are important for a variety of health- and fitness-related purposes. VO2max represents the maximal amount of oxygen an individual can transport from the environment into their lungs, diffuse into the blood, and extract at the muscles and organs to produce energy, or ATP. It represents a measure of cardiorespiratory fitness (CRF) and has a strong inverse relation with all-cause mortality and cardiovascular diseases [14–16]. VO2max also has an important relationship to endurance performance among athletes, often being cited as the most important single factor—or among the most important factors—in predicting race performance [17–19]. Pulse oximeters can non-invasively measure the amount of oxygen bound to hemoglobin based on how light reflects off the blood cells when broadcast from the device. Devices with pulse oximeters to measure BOS can also be used to monitor cardiorespiratory functions, especially in people with pulmonary diseases. It can also be useful for athletes looking to travel to altitude for an event or competition who wish to monitor their acclimatization process [20,21]. Therefore, the purpose of this study was to evaluate the accuracy (validity) of VO2max estimates and blood oxygen saturation measured via pulse oximetry using the Garmin fēnix 6 with a general population participant pool.

2. Materials and Methods

Prior to data collection occurring for this study, the protocols were approved by the University of Nevada, Las Vegas Institutional Review Board (IRB). All participants signed an informed consent and filled out pre-assessment documents prior to completing the study. While the VO2max and pulse oximetry testing were completed separately, some participants completed both and are included in each dataset. As the participant pool for both VO2max and pulse oximetry testing are different, demographic data are provided for each group.

2.1. VO2max Testing

For VO2max testing, 19 apparently healthy (people who, based on their personal knowledge, reported being healthy at the time the study was conducted), active and sedentary individuals were recruited to participate (25.50 ± 5.26 years, 11 male, 8 female, 173.63 ± 9.08 cm, 74.08 ± 14.16 kg, BMI = 24.42 ± 3.21 kg/m^2, 22.14 ± 6.06% fat mass, 36.87 ± 4.58% muscle mass, 25.07 ± 23.65 km run per week, and all reported as mean ± SD). Data collection occurred over two separate days. On the first day, participants completed a graded exercise test utilizing progressive increases in speed and grade to determine their VO2maxs. Maximal oxygen consumption was measured using the ParvoMedics TrueOne 2400 metabolic cart (ParvoMedics Inc., Salt Lake City, UT, USA). The VO2max was determined by taking the highest average oxygen consumption during the graded exercise test for a set timeframe. Aggregated VO2max values for the 4-breath, 15-s, 30-s, and 1-min averaged timeframes were obtained by the metabolic cart and served as the criteria measures for the comparisons to the WT device. The second day consisted of an outdoor run that was guided by the wearable device (Garmin fēnix 6®, Garmin Ltd., Olathe, KS, USA) to generate an estimated VO2max value. The Garmin fēnix 6 is a rugged, multisport GPS smartwatch designed for outdoor use and athletes. It is marketed to combine the functionality of a fitness tracker, outdoor navigator, and smartwatch in a durable, wrist-worn device. The participants were asked to come back between two and seven days from the first visit (5.06 ± 3.96 days). The researchers performed a factory reset on the watch prior to each subject to prevent data from previous participants from influencing the measurements and estimates of the current subject. The participants then put on the associated heart rate monitor (Garmin HRM-Run®) for the outdoor run. The outdoor run involved a 10–15 min run at an intensity above 70% of the participant's estimated max HR, according to the manufacturer's guidelines. This provided the device with enough data to estimate the VO2max, using a linear extrapolation of the heart rate (HR) and running speed [22]. The outdoor run was performed in one of two places: the University track or a flat area of the campus, depending on logistics and track availability. Five participants completed the testing at the track, and fourteen participants completed the testing on campus. The altitude was ~686 m, and the average temperature during outdoor testing was 20.67 ± 12.62 °C, as measured by local weather readings. The average distance, time, pace, and HR were 2.13 ± 0.17 km, 12.91 ± 1.42 min, 6.33 ± 1.49 min/km, and 153.50 ± 11.45 bpm, respectively, as measured by the device. The data collection took place over the timespan of ~14 months, with running trials being completed during the morning, afternoon, and evening.

2.2. Pulse Oximetry Testing

For pulse oximetry testing, 22 apparently healthy individuals were recruited to participate (25.48 ± 6.02 years, 13 male, 9 female, 173.27 ± 7.70 cm, 68.88 ± 9.10 kg, BMI = 22.91 ± 2.40 kg/m^2, 18.55 ± 7.05% fat mass, and 38.73 ± 3.61% muscle mass). The participants began by putting on the fēnix 6 on their left wrist and were instructed to have the strap tension secure but comfortable. The researchers then placed a medical-grade pulse oximeter (Roscoe Medical Fingertip Pulse Oximeter, Model: POX-ROS, Roscoe Medical Inc., Middleburg Heights, OH, USA) on the right index finger of the participant. The participants completed eight trials of testing under four conditions (two per condition). The first testing condition was under normoxic (normal oxygen concentration) conditions, with the watch head placed on the posterior wrist. The researchers performed the necessary steps (selecting the correct icon in the watch) on the watch to generate a BOS level by the fēnix 6 and recorded the value from the fingertip oximeter at the same time the watch generated a value. Afterward, the watch was then placed on the anterior wrist, and the process was repeated. After both normoxic conditions were completed, the participants performed hypoxic (low oxygen concentration) testing of the pulse oximeter. The participants were connected to an altitude simulator machine (Hypoxico Everest Summit II, Hypoxico Inc., New York, NY, USA) for a minimum of five minutes to allow for their blood oxygen levels to stabilize prior to testing. The machine was set to an altitude of 3657.6 m (12,000 ft) as the default for participants. However, if the participants became lightheaded or uncomfortable at that simulated altitude, it was lowered to an altitude better tolerated by the individual, and a five-minute waiting period reset occurred, with the possibility of returning to normoxia for as long as needed before restarting at a lower simulated altitude. All participants were seated for all pulse oximetry tests. The participants were instructed to control their breathing rate and breathed in and out in synchronization with the altitude simulator bursts of air. This corresponded to a breathing rate of 12.5 breaths per minute. Blood oxygen saturation testing under hypoxia was tested with the watch on the anterior and posterior left wrists, as was performed prior in the normoxic testing condition. The average time under hypoxia was 9.18 ± 1.05 min. If the fēnix 6 was unable to generate a measurement of BOS for any trial, the researchers retried up to three times for each trial when the watch did not generate a value on the first attempt. If it was still unable to generate a measurement after three tries, no further attempts were made. Once the values were obtained from the watch and the fingertip oximeter, the pulse oximetry testing was concluded.

2.3. Data Analysis

The VO2max values for each timeframe (4-breath, 15 s, 30 s, and 1 min) and BOS values for each condition (anterior/posterior placement, normoxia/hypoxia) were input into Google Sheets (Alphabet Inc., Mountain View, CA, USA). The pulse oximetry values were compared by condition as well as the combined dataset. All granular calculations were completed within Google Sheets. All summary statistics, validation measures, and figures were completed and generated in jamovi (jamovi Project, version 2.6.19, https://www.jamovi.org/). Descriptive statistics, error analysis (mean absolute percentage error), correlation analysis (Pearson's r, Lin's concordance correlation coefficient [CCC]), equivalence testing (TOST paired samples test), and bias assessment (Bland–Altman analysis) were also performed. The TOST test upper and lower bounds were set at +0.5 and −0.5 Cohen's D for each test. Data analysis for the VO2max data was completed by comparing the fēnix 6 estimates of the VO2max to each laboratory aggregated timeframe (4 breath, 15 s, 30 s, 1 min). Determination of validation was predetermined, and any device that produced a CCC > 0.7 and a MAPE < 10% was considered valid.

3. Results

3.1. VO2max

The 19 participants used for this analysis had an average VO2max of 48.9 mL/kg/min and an average VO2max percentile of 83.37 ± 21.14%, based on the 30 s averaged VO2max values. The error analysis showed that the fēnix 6 VO2max estimate had a MAPE of less than 10% for the 15 s, 30 s, and 1 min averaged timeframes (see Table 1). The correlation analysis produced a CCC > 0.7 for both the 15 s and 30 s averaged timeframes (see Table 1). Equivalence testing via the TOST test produced no equivalent results, with the equivalence conditions being violated for the 4-breath, 15 s, 30 s, and 1 min averaged times (see Table 1). The Bland–Altman bias values and 95% confidence intervals can be found in Table 1, and the associated plots can be found for all time parameters in Figure 1.

Table 1. VO2max descriptive and validation statistics results, n = 20. Notes: MAPE = mean absolute percentage error; TOST test = two one-sided *t*-tests. Bland–Altman bias values and 95% confidence intervals are provided. Values that met the predetermined validation criteria are bolded.

	Fēnix 6 VO2max Estimate	**Lab VO2max—4 Breath Avg**	**Lab VO2max—15 s Avg**	**Lab VO2max—30 s Avg**	**Lab VO2max—1 min Avg**
Mean (mL/kg/min)	49.68	54.54	49.95	48.94	47.91
Standard Deviation	4.61	7.28	7.04	6.67	6.76
MAPE		10.70%	**7.23%**	**7.05%**	**8.53%**
Pearson Correlation		0.73	0.78	0.78	0.76
Lin's Concordance		0.49	**0.71**	**0.73**	0.68
Bland–Altman Bias		−4.87 (−7.30, −2.44)	−0.26 (−2.45, 1.92)	0.75 (−1.28, 2.78)	1.77 (−0.35, 3.89)
TOST Test (Upper)		<0.001	0.80	0.45	0.10
TOST Test (Lower)		<0.972	0.01	0.09	0.34

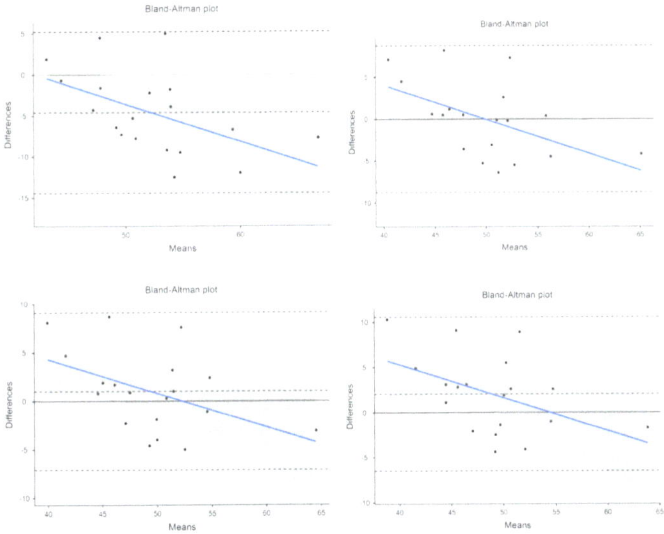

Figure 1. VO2 Bland–Altman plot of fēnix 6 compared to laboratory VO2max values: 4 s average in **top left**, 15 s average in **top right**, and 30 s average in **bottom left**, 1 min average in **bottom right**. Blue line represents proportional bias line with shadings representing 95% confidence intervals of proportional bias line. X-axis is the mean of the two measurements with the Y-axis the difference between the two measurements. The mean bias line and upper and lower limits of agreement are shown in dashed lines (mean bias being the middle-dashed line). The solid line represents the hypothetical mean bias of 0.

3.2. *Pulse Oximetry*

The error analysis showed that the fēnix 6 BOS values had a MAPE of less than 10% for all four conditions and the combined data (anterior/posterior, hypoxia/normoxia; see Table S1 and Supplementary Files). The correlation analysis did not produce a CCC > 0.7 for any conditions, including the combined data (see Table S1 and Supplementary Files). Equivalence testing via the TOST test was violated for all four conditions but was met for the combined data (see Table S1 and Supplementary Files). The Bland–Altman bias values and 95% confidence intervals can be found in Table 2 for the combined data and the Supplementary Files for individual conditions. The associated plots can be found for the combined data in Figure 2 The total number of measurements that the fēnix 6 generated was 52, for a total success rate (or data availability rate) of 59%. This means that when prompted for a blood oxygen saturation measurement, it only provided data 59% of the time.

Table 2. Blood oxygen saturation measurements measured via pulse oximetry in Garmin fēnix 6 and criterion device. Descriptive and validation statistics results for n = 22 (52 distinct fēnix 6 values from all conditions and participants). Bland–Altman bias values and 95% confidence intervals are provided. Values that met the predetermined validation criteria are bolded.

	Fēnix 6 Blood Oxygen Saturation Measurement (%)	**Criterion: Blood Oxygen Saturation Measurement (%)**
Mean	95.44%	92.06%
Standard Deviation	1.60%	8.17%
MAPE		**4.29%**
Pearson Correlation		0.18
Lin's Concordance		0.10
Bland–Altman Bias		1.12 (−0.34, 2.57)
TOST Test (Upper)		0.13
TOST Test (Lower)		0.02

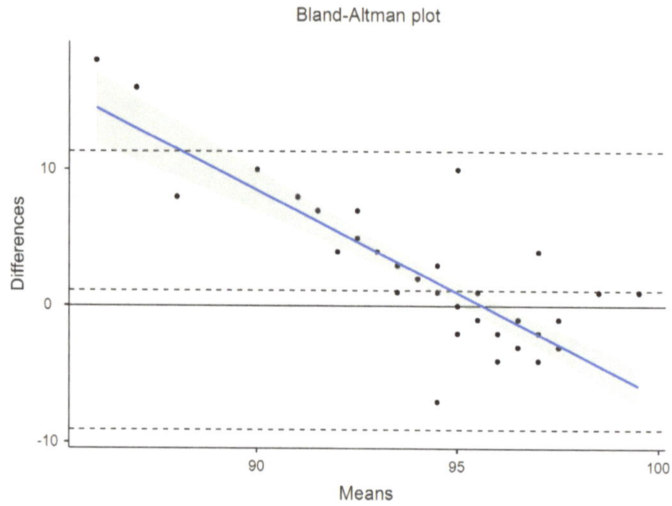

Figure 2. Bland–Altman plots for the combined pulse oximetry data, containing both hypoxia and normoxia conditions. Blue line represents proportional bias line with shadings representing 95% confidence intervals of proportional bias line. X-axis is the mean of the two measurements with the Y-axis the difference between the two measurements. The mean bias line and upper and lower limits of agreement are shown in dashed lines (mean bias being the middle-dashed line). The solid line represents the hypothetical mean bias of 0.

4. Discussion

In this study, the validity of the VO2max estimates and blood oxygen saturation (BOS) values measured via pulse oximetry in wearable technology (WT) was compared to gold-standard measurements. Based on the pre-established validation criteria, the fēnix 6 has acceptable accuracy (MAPE < 10%, CCC > 0.7) in its estimation of VO2max and corresponds closely to the 15 s and 30 s averaged timeframes. The measurements of BOS via pulse oximetry did not have acceptable accuracy for any condition or the combined data. As the appropriate use cases of these devices are discussed, it is important to note that these are consumer-grade devices, not medical devices. Thus, they are not subject to FDA regulation (or any other governing body) in terms of accuracy and effectiveness. VO2max and pulse oximeters have an important role in monitoring the health of an individual, including general health and fitness levels and those with potential cardiovascular disease (CVD) and pulmonary diseases. While these devices are being used for measuring variables in diseased populations, they are not intended for that purpose. Despite this, researchers, healthcare professionals, and public health officials are utilizing WT to track these metrics for scientific-, policy-, and healthcare-related purposes [23–29]. This illustrates the need for an independent evaluation of these devices in terms of their validity and reliability compared to gold-standard measurements. Wearable technology has the potential to revolutionize public health and physiology research due to its constant monitoring and widespread availability [12,13]. Thus, researchers, healthcare professionals, public health officials, and scientific journals should be invested in the independent validation of these devices to further several scientific fields.

Wearable technology can generate an estimate of the VO2max through the HR, as the linear relationship between the HR and VO2 is well-established [22]. The fēnix 6 measures the users HR and running speed and utilizes a linear extrapolation up to the estimated max HR, based on an individual's age, to determine the VO2max. While this can be accomplished simply with the watch and built-in photoplethysmography (PPG)-based HR monitor, an accessory HR monitor that is placed on the chest and utilizes ECG technology to determine the HR can also be used. The PPG sensors common in many watch-based wearable devices have been shown to be much less accurate at reading HRs during exercise than ECG-based HR monitors, mainly due to the PPG sensor's susceptibility to motion artifacts during movement [30–34]. ECG-based HR monitors have been recommended for use during exercise, which was observed in the current investigation. While WT represents an improvement in availability in tracking physiological metrics, such as the VO2max, field-based maximal and submaximal tests to estimate the VO2max have been around for decades [35]. A meta-analysis detailing the performance of these submaximal predictive equations compared to gold-standard testing found that they have a correlation range of r = 0.57 to 0.92 [36]. The current investigation found an r value of 0.78 for the 15 s and 30 s timeframes. Previous studies have found the Garmin fēnix 3 to have correlations of up to 0.92 [37], equal to the best submaximal equations that have been developed in terms of correlation values. Although comparing these devices solely based on correlation provides an imperfect view of their validity, accuracy, and reliability, they do offer some comparative value.

Having an accurate estimate of VO2max can be very useful, as it represents an important metric to determine a person's health status. VO2max is a reliable predictor for overall cardiorespiratory fitness (CRF), which is an independent risk factor for all-cause and disease-specific mortality [14–16]; meaning, an individual with a low VO2max value will be at a higher risk of mortality due only to that metric, regardless of any other health metrics. The American Heart Association has released a lengthy review and position statement endorsing regular measurements of CRF in clinical practice. They concluded that a substantial body of epidemiological and clinical research indicates that cardiorespiratory fitness is a potent predictor of mortality, often surpassing the predictive power of established risk factors such as smoking, hypertension, hyperlipidemia, and diabetes mellitus. Incorporating CRF into risk stratification models can substantially enhance the precision of risk assessment for adverse health outcomes [38], as an assessment of CRF is ideally performed through a maximal exercise test and measurement of oxygen consumption and carbon dioxide production through a metabolic cart. Unfortunately, this is not possible for many people who cannot complete a maximal exercise test (those with CVD, musculoskeletal diseases, pulmonary diseases, etc.) or for those who cannot afford the cost of laboratory measurements. Wearable technology has the potential to evaluate a person's VO2max through a relatively light bout of exercise (as is the case with the current device being tested) or even at rest (as is the case with other wearable devices). Thus, an accurate estimate of VO2max has the potential to influence personal health measures, as well as provide greater insights into the public health status for researchers and policymakers. As the fēnix 6 was found to generate accurate estimates of VO2max, individual recreational users, and possibly researchers, public health officials, and healthcare professionals can trust the values generated by the device. However, researchers and healthcare workers may want to utilize a more stringent validation threshold than what has been employed in the current investigation.

In addition to the role of VO2max in personal health, it is also an important measure for endurance athletes. VO2max is among the most important single measures to determine performance in an endurance event and is considered by many to be the single most important metric in determining performance [17–19]. Having the ability to know an athlete's VO2max allows for improved training programs to be developed that are tailored to the athlete's specific fitness level. As gold-standard methods of determining a person's VO2max can be expensive and time-consuming, they are not a practical option for many recreational athletes or teams. Wearable technology can represent a cost-effective method of determining aerobic capacity for individuals, as well as teams. These devices can also generate a VO2max value during the course of normal training, eliminating the need to take a day off from training for testing purposes. It also has the added benefit of constant monitoring, allowing for small changes in aerobic capacity to influence the training protocol.

Measuring BOS via pulse oximetry is a well-established and widely used method in hospitals and other clinical settings. The introduction of pulse oximetry into smartwatches and other wearable devices is a recent advancement. Pulse oximeters measure BOS by broadcasting pulses of light and measuring the reflection via PPG sensors to monitor changes in blood oxygen concentration. This technology may prove to be an important way to monitor a person's disease status and health metrics, especially those with pulmonary diseases, such as asthma, emphysema, and chronic obstructive pulmonary disease (COPD). However, independent validation of these devices will need to be completed in order to trust these measurements. It can also be useful for athletes who travel to altitude to monitor their acclimatization process, such as hikers, mountaineers, or other athletes traveling to higher altitudes than their current altitude [39]. While the device tested in the current investigation performed poorly, especially during the hypoxic conditions, it may be of interest to future researchers to test the ability of this technology to measure BOS levels accurately throughout the day rather than on demand. However, as we have mentioned previously, PPG sensors are susceptible to motion artifacts and could have similar issues with accuracy when measured throughout the day. Some research has demonstrated that desaturations below 50% can be observed when patients are moving during testing [40]. With these severe limitations in terms of the accuracy of this device, especially during hypoxic conditions, those looking to use this device to measure acclimatization when at altitude should look elsewhere for accurate measurements.

For this current investigation, we have used the generally accepted thresholds of a MAPE < 10% and CCC > 0.7. However, universal agreement for thresholds or even analytical tests to determine validity has not been established. As we recruited from the general population for this study, the fairly liberal thresholds of 10% and 0.7 seem appropriate. However, those looking to use this device in higher-level athletics, public health and/or physiology research, and healthcare may seek more conservative thresholds to determine appropriate use cases. In the future, a tiered threshold system could be established to better understand the appropriate use cases of these devices. In terms of the analytical tests, we have decided only to use MAPE and CCC in the determination of validity. However, we have also included bias assessments (Bland–Altman analysis) and equivalence testing (the TOST test). These have all been suggested as appropriate analytical techniques to determine validity, though they are not always common in other validation literature [13,41,42]. For instance, equivalence testing is especially absent from much of the validation literature. We have included all for the benefit of the reader as well as because we view them as appropriate tests to determine validity. However, because the thresholds have not been established for these additional tests, we have not included them in our validity thresholds.

Limitations

This study evaluated both active and sedentary individuals in the general population, and thus the generalizability of this device to other populations should be done cautiously, if at all. While the validation criteria used have been used in previous research, the relatively liberal thresholds (MAPE < 10% and CCC > 0.7) might not be sufficiently stringent for high-stakes applications such as high-level athletics, public health research, or healthcare settings. As this study only evaluated acute hypoxia, additional research should be performed on these devices to determine their accuracy and usefulness in monitoring blood oxygen saturation longitudinally. Finally, we have noted the environmental conditions that VO2max was tested in, as it was an outdoor running trial. As the temperature can impact a person's HR during exercise, this could be a confounding factor in the estimation. However, given that the data were collected over a ~14-month time period, this strengthens the external validity of the results and the generalizability.

5. Conclusions

In this study, we tested the Garmin fēnix 6 VO2max estimate and blood oxygen saturation values, measured via pulse oximetry for accuracy, and compared them to gold-standard laboratory measurements. The fēnix 6 showed acceptable accuracy for VO2max and was most closely aligned with the 15 s and 30 s timeframes. The fēnix 6 did not show acceptable accuracy for blood oxygen levels for any condition or the combined analysis. Therefore, the Garmin fēnix 6 may reasonably be expected to generate an accurate estimate of an individual's VO2max based on 15 s or 30 s aggregated data if more accurate laboratory tests are not available. In addition, the fēnix 6 will not generate an accurate estimate of an individual's blood oxygen levels, either in normoxia/hypoxia or utilizing anterior/posterior watch placement on the wrist.

Supplementary Materials: The following supporting information can be downloaded at: https://www.mdpi.com/article/10.3390/s25010275/s1, Table S1: Validity statistics by condition.

Author Contributions: Conceptualization, B.C. and J.W.N.; Methodology and Data Collection, B.C., S.M.C. and J.W.N.; Formal Analysis, B.C.; Original Draft Preparation, B.C.; Writing, Reviewing, and Editing, B.C., S.M.C. and J.W.N. All authors have read and agreed to the published version of the manuscript.

Funding: This research received no external funding.

Institutional Review Board Statement: This study was conducted in accordance with the Declaration of Helsinki and approved by the Institutional Review Board (or Ethics Committee) of the University of Nevada, Las Vegas (protocol code: 1525606-11; date of approval: 15 February 2023).

Informed Consent Statement: Informed consent was obtained from all subjects involved in the study.

Data Availability Statement: The data presented in this study are openly available in Excel format at https://doi.org/10.7910/DVN/43AX9B.

Conflicts of Interest: The authors declare no conflicts of interest.

References

1. Newsome, A.; Reed, R.; Sansone, J.; Batrakoulis, A.; McAvoy, C.; Parrott, M.W. 2024 ACSM Worldwide Fitness Trends: Future Directions of the Health and Fitness Industry. *ACSM's Health Fit. J.* **2024**, *28*, 14–26. [CrossRef]
2. Thompson, W.R. Worldwide Survey of Fitness Trends for 2023. *ACSM's Health Fit. J.* **2023**, *27*, 9–18. [CrossRef]
3. Thompson, W.R. Worldwide survey of fitness trends for 2022. *ACSM's Health Fit. J.* **2022**, *26*, 11–20. [CrossRef]
4. Thompson, W.R. Worldwide Survey of Fitness Trends for 2021. *ACSM's Health Fit. J.* **2021**, *25*, 10–19. [CrossRef]
5. Thompson, W.R. Worldwide survey of fitness trends for 2020. *ACSM's Health Fit. J.* **2019**, *23*, 10–18. [CrossRef]
6. Thompson, W.R. Worldwide survey of fitness trends for 2019. *ACSM's Health Fit. J.* **2018**, *22*, 10–17. [CrossRef]
7. Thompson, W.R. Worldwide survey of fitness trends for 2018: The CREP Edition. *ACSM's Health Fit. J.* **2017**, *21*, 10–19. [CrossRef]
8. Thompson, W.R. Worldwide survey of fitness trends for 2017. *ACSM's Health Fit. J.* **2016**, *20*, 8–17. [CrossRef]
9. Thompson, W.R. Worldwide survey of fitness trends for 2016. *ACSM's Health Fit. J.* **2015**, *19*, 9–18. [CrossRef]
10. Clark, E. Is Your Fitness Tracker Helping or Hurting Your Health? Available online: https://themanifest.com/app-development/fitness-tracker-helping-hurting-health (accessed on 3 August 2024).
11. Dhingra, L.S.; Aminorroaya, A.; Oikonomou, E.K.; Nargesi, A.A.; Wilson, F.P.; Krumholz, H.M.; Khera, R. Use of Wearable Devices in Individuals with or at Risk for Cardiovascular Disease in the US, 2019 to 2020. *JAMA Netw. Open* **2023**, *6*, e2316634. [CrossRef]

12. Wright, S.; Brown, T.; Collier, S.; Sandberg, K. How consumer physical activity monitors could transform human physiology research. *Am. J. Physiol. Regul. Integr. Comp. Physi* **2017**, *312*, R358–R367. [CrossRef] [PubMed]
13. Carrier, B.; Barrios, B.; Jolley, B.D.; Navalta, J.W. Validity and Reliability of Physiological Data in Applied Settings Measured by Wearable Technology: A Rapid Systematic Review. *Technologies* **2020**, *8*, 70. [CrossRef]
14. Qiu, S.; Cai, X.; Sun, Z.; Wu, T.; Schumann, U. Is estimated cardiorespiratory fitness an effective predictor for cardiovascular and all-cause mortality? A meta-analysis. *Atherosclerosis* **2021**, *330*, 22–28. [CrossRef] [PubMed]
15. Lee, D.; Sui, X.; Artero, E.G.; Lee, I.; Church, T.S.; McAuley, P.A.; Stanford, F.C.; Kohl III, H.W.; Blair, S.N. Long-term effects of changes in cardiorespiratory fitness and body mass index on all-cause and cardiovascular disease mortality in men: The Aerobics Center Longitudinal Study. *Circulation* **2011**, *124*, 2483–2490. [CrossRef]
16. Harber, M.P.; Kaminsky, L.A.; Arena, R.; Blair, S.N.; Franklin, B.A.; Myers, J.; Ross, R. Impact of cardiorespiratory fitness on all-cause and disease-specific mortality: Advances since 2009. *Prog. Cardiovasc. Dis.* **2017**, *60*, 11–20. [CrossRef]
17. Joyner, M.J.; Coyle, E.F. Endurance exercise performance: The physiology of champions. *J. Physiol.* **2008**, *586*, 35–44. [CrossRef]
18. Bassett, D.R.; Howley, E.T. Limiting factors for maximum oxygen uptake and determinants of endurance performance. *Med. Sci. Sports Exerc.* **2000**, *32*, 70–84. [CrossRef]
19. Kenney, W.L.; Wilmore, J.H.; Costill, D.L. *Physiology of Sport and Exercise*; Human Kinetics: Champaign, IL, USA, 2021.
20. Dünnwald, T.; Kienast, R.; Niederseer, D.; Burtscher, M. The use of pulse oximetry in the assessment of acclimatization to high altitude. *Sensors* **2021**, *21*, 1263. [CrossRef]
21. Sinex, J.A.; Chapman, R.F. Hypoxic training methods for improving endurance exercise performance. *J. Sport Health Sci.* **2015**, *4*, 325–332. [CrossRef]
22. Anonymous. Aerobic Fitness Level (VO2max) Estimation—Firstbeat White Paper. Available online: https://www.firstbeat.com/en/aerobic-fitness-level-vo%E2%82%82max-estimation-firstbeat-white-paper-2/ (accessed on 14 July 2022).
23. Iqbal, M.H.; Aydin, A.; Brunckhorst, O.; Dasgupta, P.; Ahmed, K. A review of wearable technology in medicine. *J. R. Soc. Med.* **2016**, *109*, 372–380. [CrossRef]
24. Park, S.; Jayaraman, S. Enhancing the quality of life through wearable technology. *IEEE Eng. Med. Biol. Mag.* **2003**, *22*, 41–48. [CrossRef] [PubMed]
25. Greiwe, J.; Nyenhuis, S.M. Wearable technology and how this can be implemented into clinical practice. *Curr. Allergy Asthma Rep.* **2020**, *20*, 1–10. [CrossRef] [PubMed]
26. Burnham, J.P.; Lu, C.; Yaeger, L.H.; Bailey, T.C.; Kollef, M.H. Using wearable technology to predict health outcomes: A literature review. *J. Am. Med. Inform. Assoc.* **2018**, *25*, 1221–1227. [CrossRef] [PubMed]
27. Qaddoori, S.L.; Fathi, I.; Hammoudy, M.A.; Ali, Q.I. Advancing Public Health Monitoring through Secure and Efficient Wearable Technology. *Int. J. Saf. Secur. Eng.* **2023**, *13*, 1001–1014. [CrossRef]
28. Phillips, S.M.; Cadmus-Bertram, L.; Rosenberg, D.; Buman, M.P.; Lynch, B.M. Wearable technology and physical activity in chronic disease: Opportunities and challenges. *Am. J. Prev. Med.* **2018**, *54*, 144–150. [CrossRef]
29. Ming, D.K.; Sangkaew, S.; Chanh, H.Q.; Nhat, P.T.; Yacoub, S.; Georgiou, P.; Holmes, A.H. Continuous physiological monitoring using wearable technology to inform individual management of infectious diseases, public health and outbreak responses. *Int. J. Infect. Dis.* **2020**, *96*, 648–654. [CrossRef]
30. Estepp, J.R.; Blackford, E.B.; Meier, C.M. Recovering pulse rate during motion artifact with a multi-imager array for non-contact imaging photoplethysmography. In Proceedings of the 2014 IEEE International Conference on Systems, Man, and Cybernetics (SMC), San Diego, CA, USA, 5–8 October 2014; IEEE: Piscataway, NJ, USA, 2014; pp. 1462–1469.
31. Lu, G.; Yang, F. Limitations of oximetry to measure heart rate variability measures. *Cardiovasc. Eng.* **2009**, *9*, 119–125. [CrossRef]
32. Carrier, B.; Salatto, R.W.; Davis, D.W.; Sertic, J.V.L.; Barrios, B.; Cater, P.; Navalta, J.W. Assessing the Validity of Several Heart Rate Monitors in Wearable Technology While Mountain Biking. *Int. J. Exerc. Sci. Conf. Proc.* **2021**, *14*, 18. [CrossRef]
33. Chow, H.; Yang, C. Accuracy of Optical Heart Rate Sensing Technology in Wearable Fitness Trackers for Young and Older Adults: Validation and Comparison Study. *JMIR mHealth uHealth* **2020**, *8*, e14707. [CrossRef]
34. Terbizan, D.J.; Dolezal, B.A.; Albano, C. Validity of seven commercially available heart rate monitors. *Meas. Phys. Educ. Exerc. Sci.* **2002**, *6*, 243–247. [CrossRef]
35. Zwiren, L.D.; Freedson, P.S.; Ward, A.; Wilke, S.; Rippe, J.M. Estimation of VO2max: A comparative analysis of five exercise tests. *Res. Q. Exerc. Sport* **1991**, *62*, 73–78. [CrossRef] [PubMed]
36. Evans, H.J.; Ferrar, K.E.; Smith, A.E.; Parfitt, G.; Eston, R.G. A systematic review of methods to predict maximal oxygen uptake from submaximal, open circuit spirometry in healthy adults. *J. Sci. Med. Sport* **2015**, *18*, 183–188. [CrossRef] [PubMed]
37. Carrier, B.; Creer, A.; Williams, L.R.; Holmes, T.M.; Jolley, B.D.; Dahl, S.; Weber, E.; Standifird, T. Validation of garmin fenix 3 HR fitness tracker biomechanics and metabolics (VO2max). *J. Meas. Phys. Behav.* **2020**, *3*, 331–337. [CrossRef]
38. Ross, R.; Blair, S.N.; Arena, R.; Church, T.S.; Després, J.; Franklin, B.A.; Haskell, W.L.; Kaminsky, L.A.; Levine, B.D.; Lavie, C.J. Importance of assessing cardiorespiratory fitness in clinical practice: A case for fitness as a clinical vital sign: A scientific statement from the American Heart Association. *Circulation* **2016**, *134*, e653–e699. [CrossRef]

39. Luks, A.M.; Swenson, E.R. Pulse oximetry at high altitude. *High Alt. Med. Biol.* **2011**, *12*, 109–119. [CrossRef]
40. Chan, E.D.; Chan, M.M.; Chan, M.M. Pulse oximetry: Understanding its basic principles facilitates appreciation of its limitations. *Respir. Med.* **2013**, *107*, 789–799. [CrossRef]
41. Welk, G.J.; Bai, Y.; Lee, J.; Godino, J.; Saint-Maurice, P.F.; Carr, L. Standardizing analytic methods and reporting in activity monitor validation studies. *Med. Sci. Sports Exerc.* **2019**, *51*, 1767. [CrossRef]
42. van Lier, H.G.; Pieterse, M.E.; Garde, A.; Postel, M.G.; de Haan, H.A.; Vollenbroek-Hutten, M.; Schraagen, J.M.; Noordzij, M.L. A standardized validity assessment protocol for physiological signals from wearable technology: Methodological underpinnings and an application to the E4 biosensor. *Behav. Res. Methods* **2020**, *52*, 607–629. [CrossRef]

sensors

MDPI

Article

Description and Classification of Training Drills, Based on Biomechanical and Physiological Load, in Elite Basketball

Carlos Sosa [1], Enrique Alonso-Pérez-Chao [2,3], Carlos Ribas [3], Xavier Schelling [4] and Alberto Lorenzo [1,*]

1 Facultad de Ciencias de la Actividad Física y del Deporte-INEF, Universidad Politécnica de Madrid, 28040 Madrid, Spain; sosamarin.carlos@gmail.com
2 Faculty of Sports Sciences, Universidad Alfonso X el Sabio, 28691 Villanueva de la Cañada, Spain; eperezch@uax.es
3 Faculty of Sports Science, Universidad Europea de Madrid, 28001 Villaviciosa de Odón, Spain; enrique.alonso@universidadeuropea.es
4 Institute for Health and Sport (iHeS), Victoria University, Melbourne 8801, Australia; xschelling@gmail.com
* Correspondence: alberto.lorenzo@upm.es

Abstract: The aim of this study was to understand and describe the physiological and biomechanical demands of various tasks used in basketball training and, subsequently, to provide a practical application of these tasks in a typical training week. Twelve basketball players had their external load variables monitored across 179 training sessions (2896 samples) using local positioning system technology. These variables included total distance covered, distance covered at various intensity levels, accelerations, decelerations, PlayerLoad™, and explosive efforts. The analysis revealed significant differences in both physiological and biomechanical loads across various drills. Specifically, tasks with more space and fewer defenders, such as 3v0 full court, impose higher physiological loads compared to tasks with less space and more defenders, like 5v5 full court. The difference in physiological load between these tasks was statistically significant ($p < 0.05$) with a moderate effect size (ES: −0.60, 95% CI: [−0.99, −0.22]). In terms of biomechanical load, drills with increased defensive pressure, such as 5v5 full court, exhibited significantly higher values compared to less specific drills, such as 5v0 full court, with a very large effect size (ES: 1.37, 95% CI: [1.04, 1.70], $p < 0.01$). Additionally, comparisons between 5v5 full court and 3v0 full court for biomechanical load produced a very large effect size (ES: 1.67, 95% CI: [1.37, 1.97], $p < 0.01$), indicating a substantial difference in load demands. The results indicate that tasks with more space and fewer defenders impose higher physiological loads, while those with less space and more defenders increase the biomechanical load. For training design, it is recommended to schedule tasks with a higher biomechanical load at the beginning of the session and those with a physiological orientation toward the end. Understanding the distinct demands of different drills can help coaches structure training sessions more effectively to optimize player load and performance development throughout the week.

Keywords: team sports; local positioning system; load monitoring; game demands; training

Academic Editors: Bruno Gonçalves, Bruno Emanuel Nogueira Figueira and Diogo Alexandre Martins Coutinho

Received: 13 November 2024
Revised: 26 December 2024
Accepted: 30 December 2024
Published: 5 January 2025

Citation: Sosa, C.; Alonso-Pérez-Chao, E.; Ribas, C.; Schelling, X.; Lorenzo, A. Description and Classification of Training Drills, Based on Biomechanical and Physiological Load, in Elite Basketball. *Sensors* **2025**, *25*, 262. https://doi.org/10.3390/s25010262

1. Introduction

In recent years, extensive research in team sports, particularly basketball, has focused on characterizing the game's physical demands [1,2], identifying the most demanding passages of play [3,4], understanding the effects of training modifications during practice sessions, and exploring injury mechanisms [5]. For instance, load quantification in basketball has been employed to analyze average and peak physical demands in competition [2,6], compare practice and competition loads [7,8], observe the evolution of training loads and performance throughout a season [9], assess the impact of targeted training programs on physical performance [10], and profiling players based on age [8,11], gender [12], position [13,14], and competitive level [15,16].

This body of research improves our understanding of basketball from a physical demands' perspective, highlighting its nature as an intermittent sport characterized by alternating offensive and defensive actions [17]. The game features frequent changes in movement type and intensity [18], with high-intensity periods interspersed with medium- and low-intensity intervals, where actions often occur unpredictably [19]. Consequently, basketball matches exhibit an irregular alternation of aerobic and anaerobic physical demands [20] that impose various neuromuscular and metabolic challenges throughout the match. Biomechanical, physiological, technical, and tactical demands in basketball create significant variability in movements and intensities, contributing to its highly intermittent nature [15,21,22].

Despite the abundance of available indicators, significant confusion and inconsistency remain in their application and integration into training processes. The current basketball literature is often more descriptive than practical and offers limited information on physiological or biomechanical stressors. As Russell et al. (2020) noted in their systematic review, there is a clear disconnect between applied practices and methodological frameworks. Given the limitations of existing studies, drawing definitive conclusions about the true physical demands of basketball is not yet possible. Most research to date (e.g., [4,5,23]) has analyzed the loads experienced by basketball players using the traditional distinction between external and internal loading. However, several studies have highlighted the challenges and critical importance of distinguishing between physiological and biomechanical load–response pathways in team sports, including basketball [23].

Monitoring physiological and biomechanical load adaptation in team sports, particularly basketball, has gained increasing recognition and research interest. Although physiological loads primarily focus on the work–energy relationship as players move around the court, biomechanical loads refer to the external forces exerted on the players through their movements. Biomechanical loads refer to the external forces exerted on players through their movements, including the impact of gravity, ground reaction forces, resistance from equipment, and interactions with opponents. These loads include the stresses placed on muscles, tendons, bones, and joints during physical activities, which influences both performance and the risk of injury [24]. It is well established that biomechanical and neuromuscular factors play a critical role in horizontal deceleration, a key component of sports that involve multidirectional movements, such as basketball. The unique ground reaction force profile during horizontal deceleration, characterized by high-impact peak forces and loading rates, can increase susceptibility to excessive forces and the risk of injury if the limbs are unable to tolerate these forces [25]. Various metrics are available to quantify the biomechanical loads experienced by the body, its structures, and individual tissues. However, the challenge lies in measuring these loads accurately, both in and out of laboratory settings [26]. The biomechanical load, defined as the external forces acting on an athlete's body, can be monitored using Inertial Measurement Units (IMUs). These devices utilize inertial movement analysis (IMA) to capture biomechanical load variables such as player load or jumps. IMUs are widely used to monitor adaptations to training and their relationship with game performance.

For example, in football, Mandorino et al. [27] employed machine learning techniques to develop a novel locomotor efficiency index (LEI) to assess the neuromuscular fitness of players. Subsequently, Mandorino et al. [28] analyzed the effects of different training periodization strategies on the neuromuscular state of football players. In basketball, a study quantifying workload during basketball-specific drills using microtechnology revealed that full-court 3v3 and 5v5 drills imposed the highest physical demands compared to traditional balanced basketball drills such as 2v2 and 4v4. Metrics such as acceleration load per minute (AL·min^{-1}) were used to assess workload, demonstrating that the drill format significantly influences the biomechanical load experienced by players [29].

More recently, Olthof et al. (2021) studied the statistical relationships between biomechanical loads in training and game performance. Their findings indicated that training loads significantly affected match loads in subsequent games. In particular, increasing training loads two days before a match led to higher expected match loads, suggesting that biomechanical loads are strong predictors of game performance.

What is clear so far is that manipulating certain variables, such as the number of players involved in the use of full-court versus half-court drills, results in significantly different player loads [2,30]. Considering these findings, the biomechanical load of various basketball exercises must be considered, with the overarching goal of training being to prepare players for competition. Training sessions should be designed to reflect this goal by observing the distinct loads induced by each exercise, not only in terms of internal and external load, but also in relation to physiological and biomechanical load. By understanding the specific physiological and biomechanical demands, internal or external, of each task assigned to players, the designs of the training sessions can be optimized, ultimately leading to improved performance in competition. Addressing the gap between theoretical knowledge and practical application is crucial because it ensures that research findings are not confined to academic settings but are effectively translated into real-world practice. In the context of basketball training, presenting actionable strategies enables coaches and practitioners to directly apply evidence-based insights to their training designs. This not only enhances the relevance and utility of the research but also helps optimize training effectiveness, ensuring that players are better prepared for the demands of competition.

To our knowledge, there is limited research that specifically differentiates between the types of loads (physiological and/or biomechanical) for each task in basketball. Furthermore, even fewer studies go beyond describing the load and offer practical applications of this knowledge for improved training design. Thus, the aim of this article is twofold; first, to identify and describe the physiological and biomechanical load associated with various tasks used in basketball training and, second, to propose a practical application of these tasks within the framework of a typical training week.

2. Materials and Methods

2.1. Sample

Elite male basketball players [31] from the same team competing in the highest regional division of an U18 Spanish basketball competition were included in this study (n = 18, mean ± standard deviation [SD]: age 16.9 ± 0.8 years, height 196.6 ± 9.4 cm, body mass: 91.7 ± 8.2 kg). Monitoring took place during 179 training sessions.

Data collection was carried out at the same facility for two consecutive seasons (2018–2019 and 2019–2020). To be included in the study, players had to complete a minimum of 50% of training sessions (n = 90/179) during both seasons; those who did not meet this criterion were excluded from the analysis. Additionally, data from players who did not complete at least 80% of the total duration of a specific training session were excluded from that session's data pool but remained in the overarching study.

After applying the exclusion criteria, six participants who entered the study were excluded from the analysis. Consequently, 2896 training data samples from a collective of

12 participants were subjected to analysis. This study was conducted in accordance to the Declaration of Helsinki [32].

2.2. Procedures

This observational investigation was conducted over a 2-year period throughout the 2019–2020 and 2020–2021 seasons. Each player wore a device (Vector S7; Catapult Sports, Melbourne, Australia) in a specially designed pocket within a vest, placed on the upper thoracic spine between the scapulae. The devices contained an accelerometer (±16 g, 100 Hz), magnetometer (±4900 µT, 100 Hz), gyroscope (up to 2000 deg/s, 100 Hz), and LPS. The ClearSky LPS (ClearSky S7, 10 Hz, firmware version 5.6.; Catapult Sports, Melbourne, Australia) is an ultra-wide band, 4 GHz transmitting system equipped with 24 anchors positioned around the perimeter of a basketball stadium that was used to collect LPS data. The technology used in this study has been supported as valid in measuring distance [33–36], speed, accelerations, decelerations [33,34], and Player Load^{TM} [37], while similar LPS technology has been shown to be reliable (coefficient of variation (CV) < 5%) in measuring distance and speed variables [36]. All players were familiar with monitoring technology, having worn the devices during training and games in the previous season. Each device was turned on ~20–40 min before the warm-up that preceded each game. The players wore the same device throughout the study period to avoid variation between devices in the output of external load data [38].

Activity editing occurred during and after the session. To minimize significant interobserver variability, the editing process for all activities was carried out consistently by the same individual. During training sessions, duration was defined as the time in minutes that a player actively participated in training, excluding the intervals between exercises, hydration breaks, or instances when a player, during a task, was not actively involved. A player was considered inactive during a task if they were off the court and did not participate (e.g., in a 5v5, where a player awaits off-court to substitute for a teammate). After completing data collection, Catapult Sports Openfield cloud software (version 1.22.0) was used to extract data from each player for each training session, segmented by task. Subsequently, following the predefined exclusion criteria, the collected data were exported into a Microsoft Excel spreadsheet (version 16.0, Microsoft Corporation, Redmond, WA) for further analysis. The drills were classified according to their specificity from 0 to 5, following the classification by [29]. Activities at level 0–1 were those carried out outside the basketball court and unrelated to basketball practice (e.g., cycling), while level 5 represented an official basketball game (Table 1).

Table 1. Drill classification based on their specificity from 5 to 0 (Schelling & Torres, 2016).

4	5v5 full court	A 5v5 game is played with 10 players on the court at the same time. The number of consecutive plays and the work-to-rest ratio vary depending on the coach's feedback and the pauses they implement.	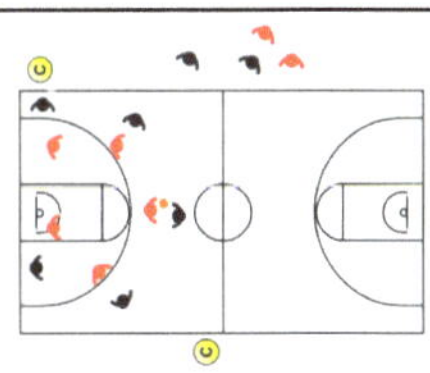
4	5v5v5	A 5v5v5 game is played where the defending team transitions to offense and attacks the opposite basket.	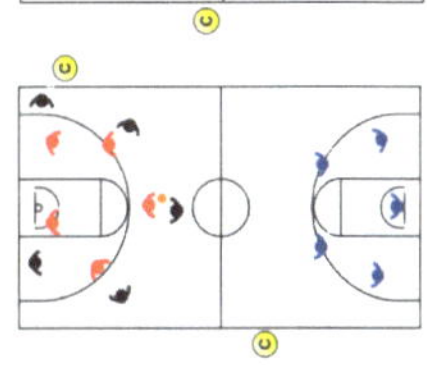

Table 1. *Cont.*

4	4v4 full court	A 4v4 game is played with 8 players on the court at the same time. When the offensive play ends, the defending team transitions to offense and attacks the same team at the opposite basket. The number of consecutive plays and the work-to-rest ratio ranges between 3 and 6, depending on the coach's feedback and pauses	
4	4v4v4	A 4v4v4 game is played where the defending team transitions to offense and attacks the opposite basket, where another team is waiting to defend.	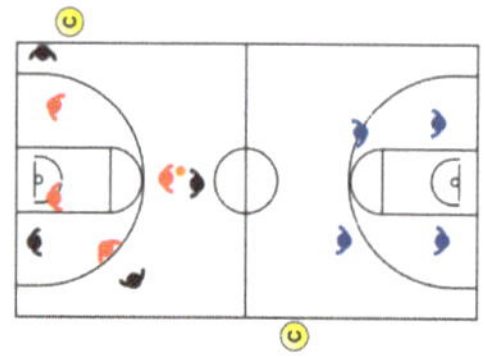
3	3v3 Full court	A 3v3 game is played with 6 players on the court at the same time. When the offensive play ends, the defending team transitions to offense and attacks the same team at the opposite basket. The number of consecutive plays and the work to rest ratio vary between 3 and 6, depending on the coach's feedback and pauses.	
3	3v3v3	A 3v3v3 game is played where the defending team transitions to offense and attacks the opposite basket, where another team is waiting to defend.	
3	Eleven Player Break	A continuous 3v2 situation is played. Among the five players involved, the one who gains possession when the play ends (whether through a basket, rebound, or turnover) attacks on the opposite side with two players positioned in the corners against two defenders waiting on the other side.	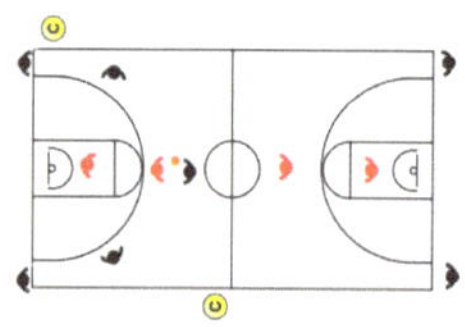
3	2v2 full court	A 2v2 game is played where, after an offensive play, the team defends at the opposite basket. Following the defensive effort, the team passes to one of the two teammates positioned to transition and attack on the opposite court.	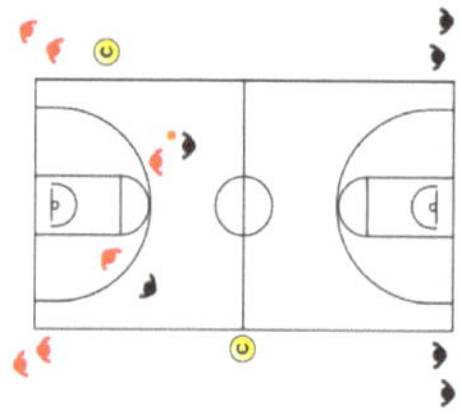
2	1v1 in longitudinal half court (28 × 7.5 m)	The attacking player must attempt to drive past and score after playing a 1v1. Once the offensive play is over, the player who attacked transitions to defense.	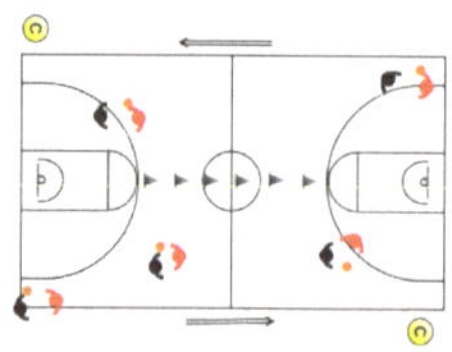
2	5v0 full court	A 5v0 drill is performed at midcourt, followed by another drill at the opposite end. After completing these drills, 5 new players enter the court.	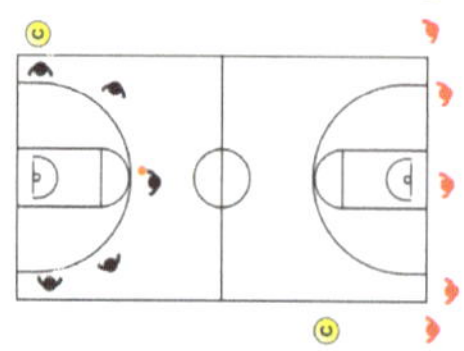

Table 1. *Cont.*

2	4v0 full court	A 4v0 drill is performed at midcourt, followed by another drill at the opposite end. After completing these drills, 4 new players enter the court.	
2	3v0 full court	A 3v0 drill is performed at midcourt, followed by another drill at the opposite end. After completing these drills, 3 new players enter the court.	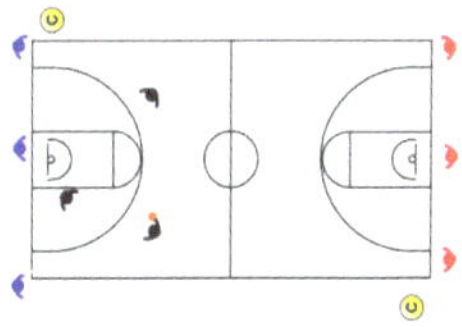
2	2v0 (Individual Technical-Tactical) half court	Different individual technical-tactical situations are practiced without opposition in a 2v0 setting	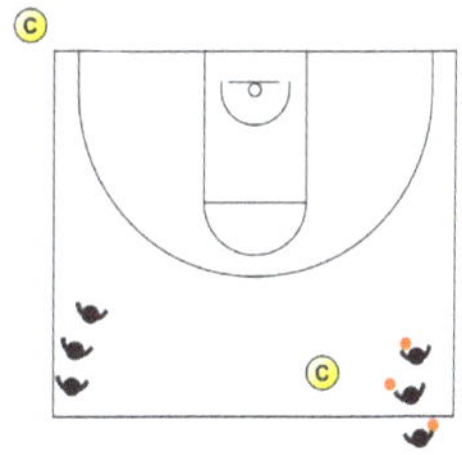
2	1v0 (Individual Technical-Tactical) half court	Different individual technical-tactical situations are practiced without opposition in a 1v0 setting.	

2.3. Physical Variables

The selected physical parameters were classified into two types (physiological and biomechanical variables) [23,39]. Each variable was extracted and represented as a relative value, indicating the rate of accumulation of that parameter per minute.

2.3.1. Physiological Variables

The following 5 variables were considered physiological: distance (m) per minute covered (TD) and distance (m) per minute covered in different intensity zones including: standing–walking (S-W) = <7 km·h^{-1}; jogging (JOG) = 7–14 km·h^{-1}; running (RUN) = 14.01–18 km·h^{-1}; and high-speed running (HSR) >18 km·h^{-1}, as previously used in basketball research [15].

To classify tasks according to the orientation of the physiological load, a two-step cluster analysis was performed (average silhouette = 0.5) using physiological parameters: total distance per minute, and distance per minute at different thresholds (Table 2). Tasks were grouped into four categories: low physiological load, medium physiological load, high physiological load, and very high physiological load. Each category was assigned a numerical value, with 1 representing low physiological load, 2 for medium physiological load, 3 for high physiological load, and 4 for very high physiological load.

Table 2. Cluster analysis identifying drill groups based on physiological load parameters.

Variables	Physiological Load			
	Low	Medium	High	Very High
Distance per minute (m)	18.56	62.09	75.28	80.50
Standing–walking (<7 km·h^{-1})	15.53	31.75	65.40	34.35
Jogging (7–14 km·h^{-1})	3.13	21.36	51.81	28.05
Running (14.01–18 km·h^{-1})	0.85	6.37	17.32	12.29
High-speed running (>18 km·h^{-1})	0.22	2.42	3.77	6.02
Sample size (N)	141	1831	122	1042
Sample proportion (%)	4.5%	58.4%	3.9%	32.2%
Bayesian information criterion (BIC)	9214.44			
Average silhouette	0.5			

Note: The value of each physiological load variable is presented as the mean and standard deviation for each group, and the sample size indicates the number of tasks included in each group.

To determine the physiological load of each task, an average was calculated for each task based on the numerical value of the cluster load ranging from 1 to 4. For instance, if 100 official match records were distributed with 50 in cluster number 4 and 50 in cluster number 3, the average of the 100 records would be a value of 3.5, indicating a physiological load of 3.5.

2.3.2. Biomechanical Variables

The following 5 variables were considered biomechanical: jumps per minute (JUMPS) > 20 cm, accelerations per minute (ACC) (count) performed > 2 m·s^{-2} (dwell time: 0.3 s), decelerations per minute (DEC) (count) performed > −2 m·s^{-2} (dwell time: 0.3 s), PlayerLoad™ per minute (PL) (arbitrary units [AU]), and explosive efforts per minute (EE). These dwell times were chosen given values between 0.3 and 0.4 s have been identified as the most readily used in basketball settings [40–42].

PL was calculated as the square root of the sum of the instantaneous rate of change in acceleration in the three movement planes (x-, y-, and z-axes) using the following formula [6,40]: $PlayerLoad^{TM} = [\sqrt{(a_{y1} - a_{y-1})}^2 + \sqrt{(a_{x1} - a_{x-1})}^2 + \sqrt{(a_{z1} - a_{z-1})}^2]/100$, where fwd indicates movement in the anterior–posterior direction, side indicates movement in the medial–lateral direction, up indicates vertical movement, and t represents time, while EE was calculated as the number of inertial movements per minute (n·min) derived from the analysis of high- and medium-intensity inertial movements (accelerations, decelerations, and changes of direction).

To group the tasks based on the biomechanical load orientation, a two-step cluster analysis was conducted (average silhouette = 0.5) using the biomechanical parameters JUMPS, ACC, DEC, PL, and EE (Table 3). Exercises were grouped into tasks with low biomechanical load and tasks with high biomechanical load. A numerical value of 1 was assigned to low biomechanical load, while a value of 2 was assigned to high biomechanical load. To determine the biomechanical load of each task, an average was calculated for each task based on the numerical value of the cluster load ranging from 1 to 2.

Table 3. Cluster analysis identifying drill groups based on biomechanical load parameters.

Variables	Biomechanical Load	
	Low	High
Accelerations per minute	1.35	2.71
Decelerations per minute	1.20	3.38

Table 3. *Cont.*

Variables	Biomechanical Load	
	Low	High
Explosive efforts per minute	1.56	3.26
PlayerLoad per minute	5.91	8.42
Jumps per minute	0.65	0.73
Sample size (N)	128	2124
Sample proportion (%)	47.4%	67.7%
Bayesian Information Criterion (BIC)	10,677.49	
Average silhouette	0.5	

Notes: The value of each physiological load variable is presented as the mean and standard deviation of each group, and the sample size indicates the number of tasks included in each group.

2.4. Statistical Analysis

The mean, standard deviation (SD), and coefficient of variation (CV) were determined to describe the external physical load for each drill, while to describe the load orientation, the mean, median, and SD were utilized.

A Linear Mixed Model (LMM) was used to identify differences in external load and its orientation between drills (1v0-Individual Technical-Tactical-half court, 1v1 in longitudinal half court-28 × 7.5 m-, 2v0-Individual Technical-Tactical-half court, 2v2 full court, 3v0 full court, 3v3 full court, 3v3v3, 4v0 full court, 4v4 full court, 4v4v4, 5v0 full court, 5v5 full court, 5v5v5, and Eleven player break).

"Player" was used as a random effect. Tasks were included as nominal predictor variables in the LMM at 14 levels (1v0-Individual Technical-Tactical-half court, 1v1 full court in longitudinal half-28 × 7.5 m-, 2v0-Individual Technical-Tactical-half court, 2v2 full court, 3v0 full court, 3v3 full court, 3v3v3, 4v0 full court, 4v4 full court, 4v4v4, 5v0 full court, 5v5 full court, 5v5v5, Eleven player break).

Cohen's effect size (ES) and the mean difference with 95% confidence intervals (CI) were determined for all pairwise comparisons and interpreted as follows: trivial = <0.20; small = 0.20–0.59; moderate = 0.60–1.19; large = 1.20–1.99; and very large = ≥2.00 [43]. All analyses were conducted using IBM SPSS for Windows (version 23, IBM Corporation, Armonk, NY, USA), except ES, which was calculated using a customized Microsoft Excel spreadsheet (version 16.0, Microsoft Corporation, Redmond, WA, USA).

3. Results

The descriptive analysis of each drill according to physical orientation (physiological or biomechanical) and specificity is presented in Table 4. The distribution of drills based on the orientation of the training load orientation is shown in Figure 1.

Table 4. Descriptive statistics for each drill according to physical orientation (physiological or biomechanical) and specificity.

Drill	Specificity	Physiological Load		Biomechanical Load	
		Median	Mean ± SD (% CV)	Median	Mean ± SD (% CV)
5v5 full court	4	2	2.84 ± 0.99 (35%)	2	1.79 ± 0.40 (22%)
5v5v5		2	2.00 ± 0.00 (0%)	2	1.64 ± 0.48 (29%)
4v4 full court	3	2	2.96 ± 1.01 (34%)	2	1.82 ± 0.38 (21%)
4v4v4		2	2.11 ± 0.51 (24%)	2	1.62 ± 0.48 (30%)

Table 4. *Cont.*

Drill	Specificity	Physiological Load		Biomechanical Load	
		Median	Mean ± SD (% CV)	Median	Mean ± SD (% CV)
3v3 Full court	**3**	4	3.11 ± 0.99 (32%)	2	1.82 ± 0.38 (21%)
3v3v3		2	2.34 ± 0.75 (32%)	2	1.75 ± 0.43 (25%)
Eleven Player Break	**2**	4	3.06 ± 1.00 (33%)	2	1.64 ± 0.48 (29%)
2v2 full court		2	2.59 ± 0.99 (38%)	2	1.62 ± 0.48 (30%)
1v1 in longitudinal half court (28 × 7.5 m)		2	2.20 ± 0.62 (28%)	2	1.83 ± 0.37 (20%)
5v0 full court		2	2.54 ± 0.92 (36%)	1	1.32 ± 0.46 (35%)
4v0 full court		2	2.42 ± 0.81 (33%)	1	1.24 ± 0.43 (35%)
3v0 full court		3	3.03 ± 0.97 (32%)	1	1.13 ± 0.34 (30%)
2v0 (Individual Technical-Tactical) half court		2	2.34 ± 0.72 (31%)	1	1.17 ± 0.38 (32%)
1v0 (Individual Technical-Tactical) half court		2	1.94 ± 0.24 (12%)	1	1.21 ± 0.40 (33%)

Figure 1. Distribution of drills based physical orientation (physiological or biomechanical) and specificity. *Note:* The colors are related to the specificity of the exercises, as shown in Table 4, evolving from less specific exercises -blue- to more specific exercises -red-.

The descriptive analysis (mean ± SD, and % CV) of the external physical load of training drills and the effect size ± 95% CI of the differences between tasks (1v0-Individual Tactical-Technical-half-court vs. 1v1 full court in longitudinal middle-28 × 7.5 m-, 2v0-Individual Tactical-Technical-half-court, 2v2 full court, 3v0 full court, 3v3 full court, 3v3v3, 4v0 full court, 4v4 full court, 4v4v4, 5v0 full court, 5v5 full court, 5v5v5, Eleven Player Break) are shown in Figure 2.

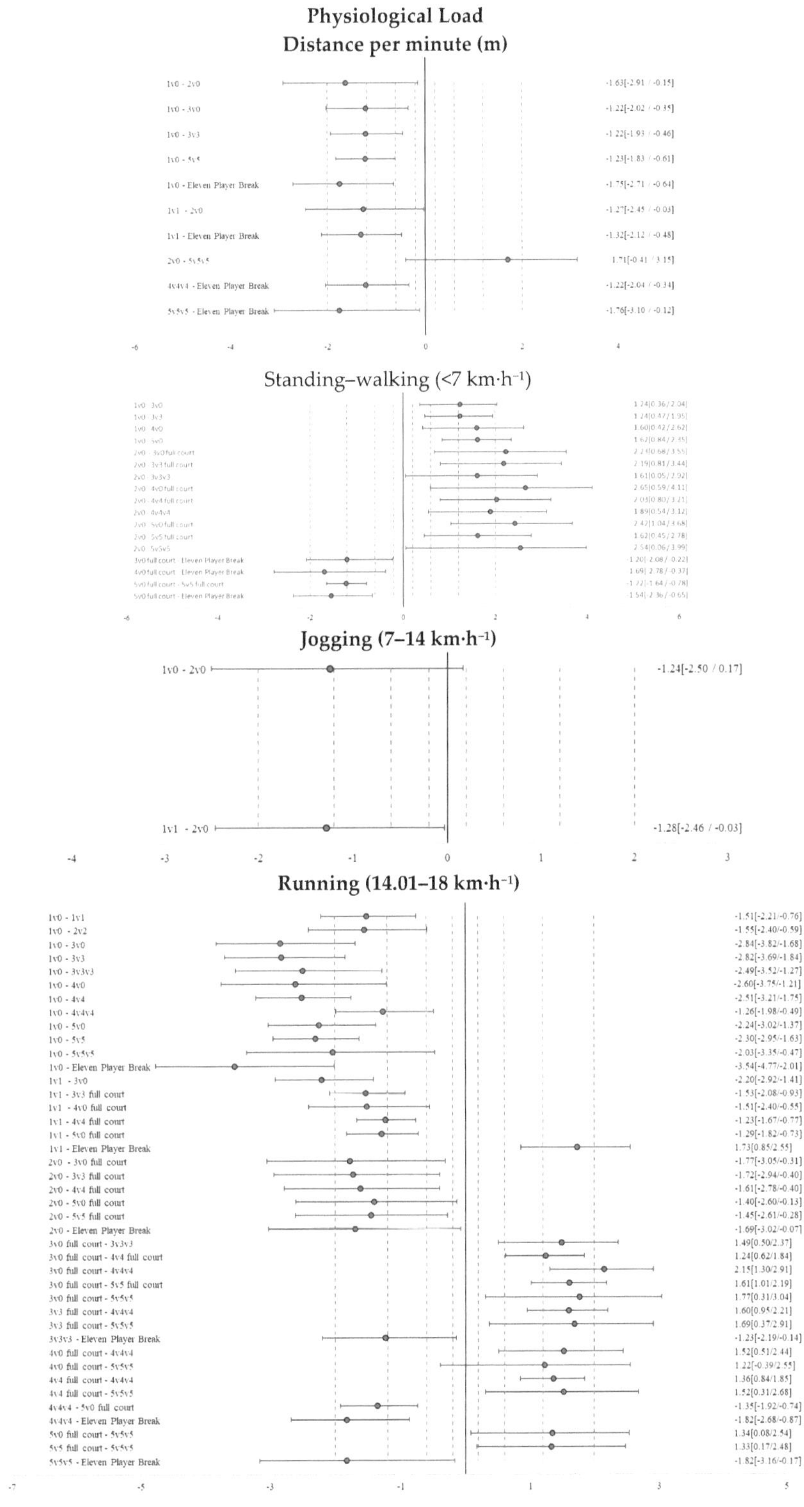

Figure 2. *Cont.*

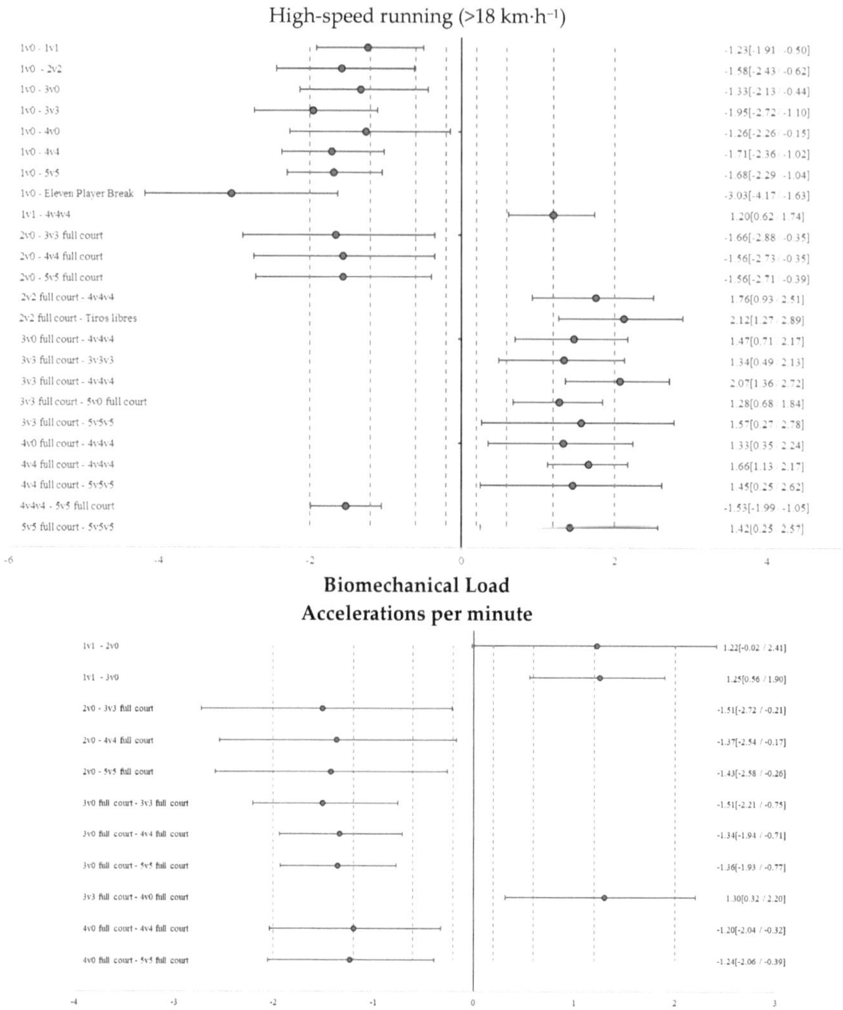

Figure 2. *Cont.*

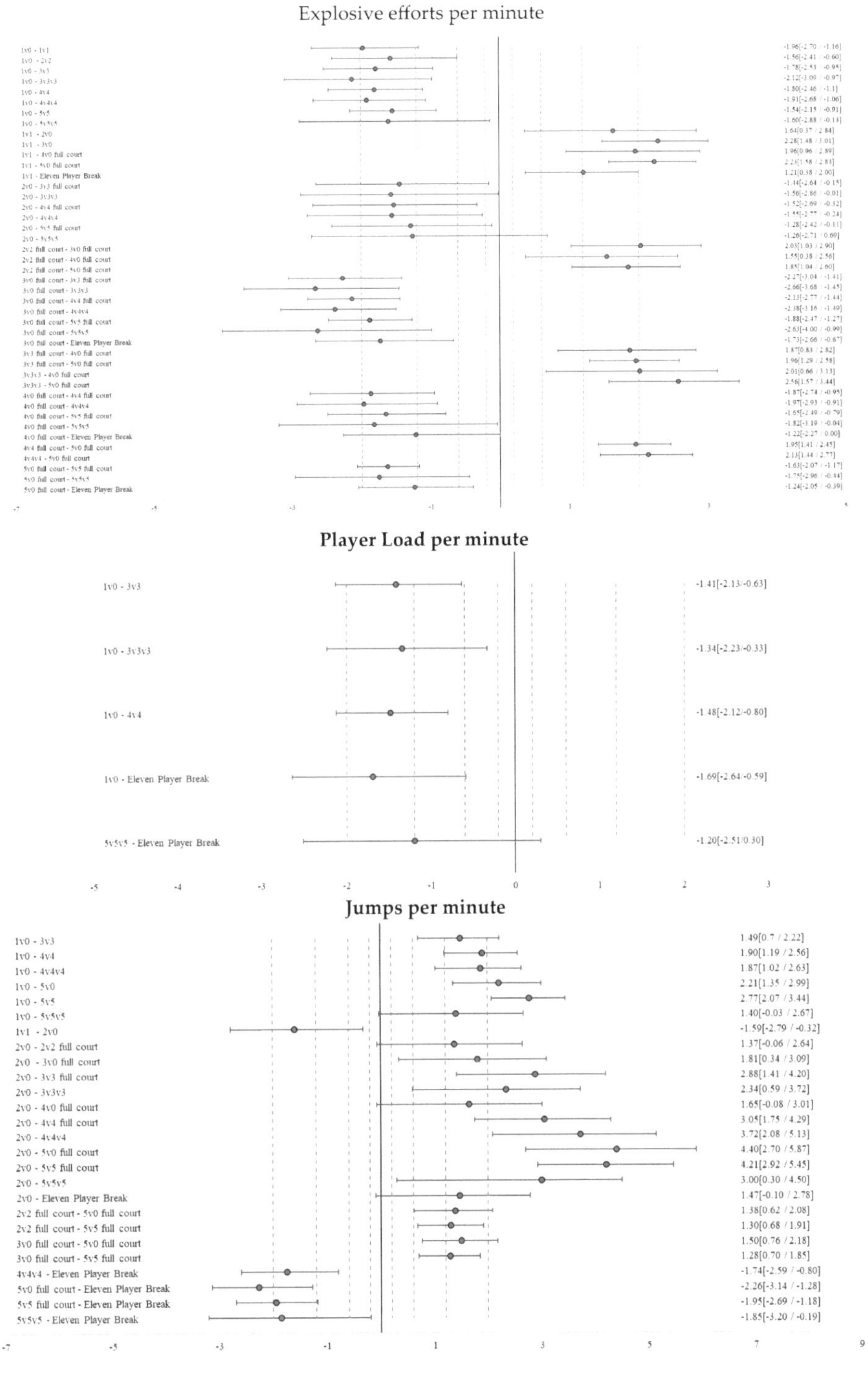

Figure 2. Standardized differences (Cohen's d) and their respective 95% confidence intervals (CI) between the training tasks that showed significant large–very large size differences for physiological and biomechanical loads.

Regarding the comparison for physiological load (Figure 3), it is notable that 4v4v4 was significantly lower than 3v3 full court (ES: −1.26), Eleven Player Fast Break (ES: −1.45), and 3v0 full court (ES: −1.30). Furthermore, Eleven Player Fast Break showed significantly higher values than 1v1 full court (ES: 1.23).

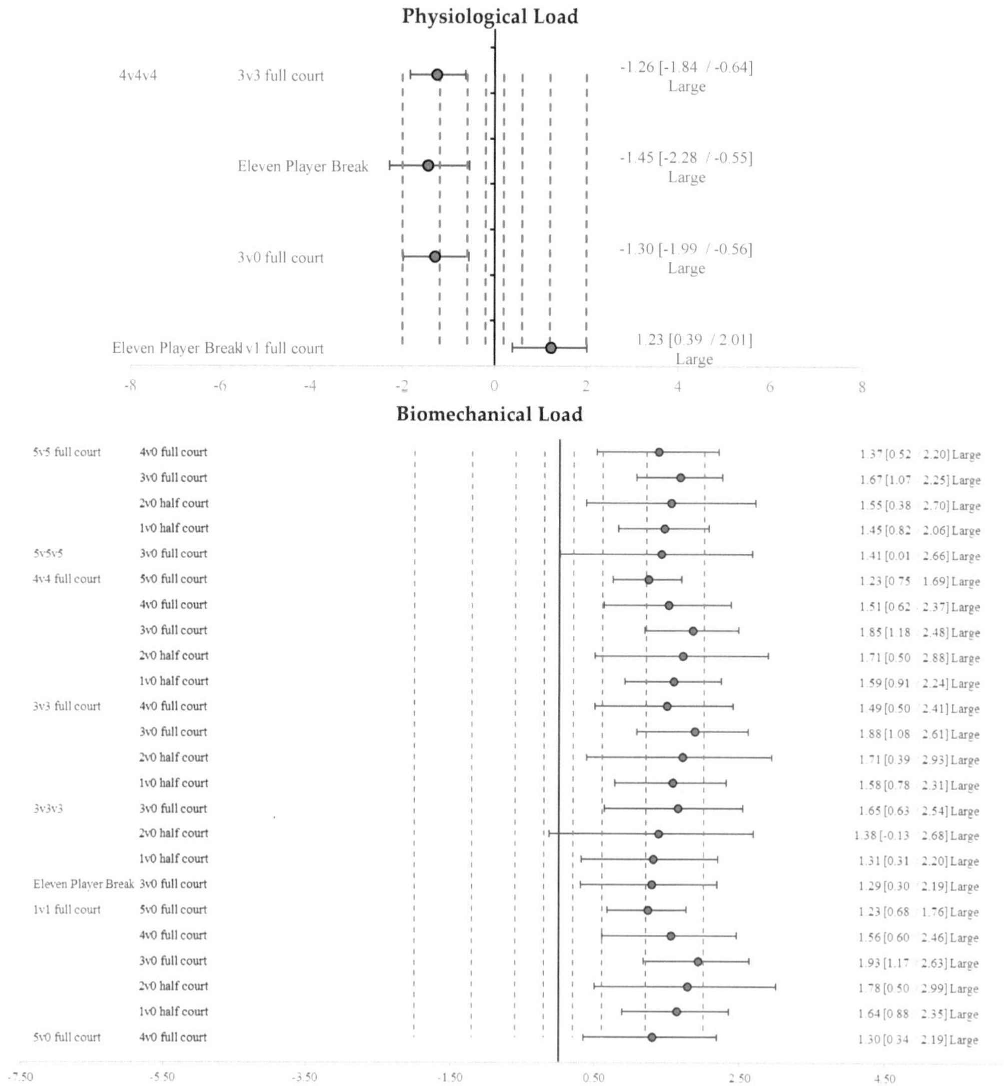

Figure 3. Standardized differences (Cohen's d) and their respective 95% confidence intervals (CI) between training tasks and match tasks for physiological and biomechanical load. Notes: The dashed line represents the magnitude of the effect from large to very large.

Regarding the comparison between tasks for the biomechanical load (Figure 3), 5v5 full court, 4v4 full court, and 3v3 full court showed significantly higher values than 4v0 full court (ES 5v5 full court: 1.37; ES 4v4 full court: 1.23; ES 3v3 full court: 1.49), 3v0 full court (ES 5v5 full court: 1.67; ES 4v4 full court: 1.85; ES 3v3 full court: 1.88), 2v0 half court (ES 5v5 full court: 1.71; ES 4v4 full court: 1.71; ES 3v3 full court: 1.38), and 1v0 half court (ES

5v5 full court: 1.59; ES 4v4 full court: 1.58; ES 3v3 full court: 1.31). Additionally, 4v4 full court also showed significantly higher values than 5v0 full court (ES: 1.23).

Concerning 3v3v3, the results showed significantly higher values than 3v0 full court (ES: 1.65), 2v0 half court (ES: 1.38), and 1v0 half court (ES: 1.31). Moreover, Eleven Player Fast Break reached significantly higher values than 3v0 full court (ES: 1.29). Regarding 1v1 full court, it showed significantly higher values compared to 5v0 full court (ES: 1.23), 4v0 full court (ES: 1.56), 3v0 full court (ES: 1.93), 2v0 half court (ES: 1.78), and 1v0 half court (ES: 1.64). Finally, 5v0 full court obtained significantly higher values than 4v0 full court (ES: 1.30).

Regarding the comparisons of physiological and biomechanical load, significant differences ($p < 0.05$) with effect sizes ranging from trivial to very large are shown in Table 5.

Table 5. Comparisons for each drill according to physiological or biomechanical orientation.

		Physiological Load		Biomechanical Load	
		Dif. Mean [I/S]	**Sig.**	**Dif. Mean [I/S]**	**Sig.**
1v0	*1v1*	−0.26 [−0.63/0.10]	1.000	−0.62 * [−0.79/−0.45]	0.000
	2v0	−0.41 [−1.06/0.25]	1.000	0.04 [−0.26/0.34]	1.000
	2v2	−0.64 * [−1.09/−0.21]	0.000	−0.41 * [−0.61/−0.2]	0.000
	3v0	−1.09 * [−1.54/−0.65]	0.000	0.07 [−0.13/0.28]	1.000
	3v3	−1.16 * [−1.54/−0.80]	0.000	−0.61 * [−0.78/−0.44]	0.000
	3v3v3	−0.40 [−0.91/0.11]	0.670	−0.54 * [−0.77/−0.3]	0.000
	4v0	−0.48 [−1.02/0.06]	0.190	−0.03 [−0.28/0.22]	1.000
	4v4	−1.02 * [−1.37/−0.68]	0.000	−0.61 * [−0.77/−0.45]	0.000
	4v4v4	−0.17 [−0.56/0.21]	1.000	−0.41 * [−0.59/−0.24]	0.000
	5v0	−0.60 * [−0.99/−0.22]	0.000	−0.11 [−0.28/0.07]	1.000
	5v5	−0.90 * [−1.23/−0.57]	0.000	−0.59 * [−0.74/−0.43]	0.000
	5v5v5	−0.06 [−0.65/0.52]	1.000	−0.43 * [−0.7/−0.16]	0.000
	Eleven Player Break	−1.13 * [−1.68/−0.58]	0.000	−0.43 * [−0.68/−0.18]	0.000
1v1	*2v0*	−0.14 [−0.75/0.46]	1.000	0.65 * [0.38/0.93]	0.000
	2v2	−0.38 * [−0.75/−0.02]	0.020	0.21 * [0.05/0.38]	0.000
	3v0	−0.83 * [−1.19/−0.47]	0.000	0.69 * [0.53/0.86]	0.000
	3v3	−0.90 * [−1.17/−0.64]	0.000	0.01 [−0.11/0.13]	1.000
	3v3v3	−0.14 [−0.58/0.31]	1.000	0.08 [−0.12/0.28]	1.000
	4v0	−0.22 [−0.69/0.26]	1.000	0.59 * [0.37/0.8]	0.000
	4v4	−0.76 * [−0.99/−0.53]	0.000	0.01 [−0.09/0.11]	1.000
	4v4v4	0.09 [−0.19/0.37]	1.000	0.21 * [0.08/0.34]	0.000
	5v0	−0.34 * [−0.62/−0.05]	0.000	0.51 * [0.38/0.64]	0.000
	5v5	−0.64 * [−0.85/−0.43]	0.000	0.03 [−0.06/0.13]	1.000
	5v5v5	0.20 [−0.33/0.73]	1.000	0.19 [−0.06/0.43]	0.800
	Eleven Player Break	−0.86 * [−1.35/−0.38]	0.000	0.19 [−0.03/0.41]	0.340

Table 5. *Cont.*

		Physiological Load		Biomechanical Load	
		Dif. Mean [I/S]	Sig.	Dif. Mean [I/S]	Sig.
2v0	*2v2*	−0.24 [−0.89/0.41]	1.000	−0.44 * [−0.74/−0.14]	0.000
	3v0	−0.69 * [−1.34/−0.03]	0.030	0.04 [−0.26/0.34]	1.000
	3v3	−0.76 * [−1.37/−0.16]	0.000	−0.64 * [−0.92/−0.37]	0.000
	3v3v3	0.01 [−0.70/0.71]	1.000	−0.57 * [−0.89/−0.25]	0.000
	4v0	−0.08 [−0.80/0.65]	1.000	−0.07 [−0.4/0.26]	1.000
	4v4	−0.62 * [−1.21/−0.03]	0.030	−0.65 * [−0.91/−0.37]	0.000
	4v4v4	0.23 [−0.38/0.85]	1.000	−0.45 * [−0.73/−0.17]	0.000
	5v0	−0.20 [−0.81/0.42]	1.000	−0.14 [−0.42/0.14]	1.000
	5v5	−0.49 [−1.08/0.09]	0.340	−0.62 * [−0.89/−0.36]	0.000
	5v5v5	0.34 [−0.41/1.10]	1.000	−0.47 * [−0.82/−0.12]	0.000
	Eleven Player Break	−0.72 [−1.45/0.01]	0.060	−0.47 * [−0.8/−0.13]	0.000
2v2	*3v0*	−0.45 * [−0.89/0.00]	0.050	0.48 * [0.28/0.68]	0.000
	3v3	−0.52 * [−0.88/−0.16]	0.000	−0.20 * [−0.37/−0.03]	0.000
	3v3v3	0.25 [−0.26/0.76]	1.000	−0.13 [−0.36/0.1]	1.000
	4v0	0.17 [−0.37/0.70]	1.000	0.38 * [0.13/0.62]	0.000
	4v4	−0.38 * [−0.71/−0.04]	0.010	−0.20 * [−0.36/−0.05]	0.000
	4v4v4	0.47 * [0.10/0.85]	0.000	0 [−0.18/0.17]	1.000
	5v0	0.05 [−0.33/0.43]	1.000	0.30 * [0.13/0.48]	0.000
	5v5	−0.25 [−0.58/0.08]	0.790	−0.18 * [−0.33/−0.03]	0.000
	5v5v5	0.59 * [0.00/1.17]	0.050	−0.02 [−0.29/0.24]	1.000
	Eleven Player Break	−0.48 [−1.03/0.07]	0.250	−0.02 [−0.27/0.23]	1.000
3v0	*3v3*	−0.07 [−0.44/0.29]	1.000	−0.68 * [−0.85/−0.51]	0.000
	3v3v3	0.69 * [0.18/1.20]	0.000	−0.61 * [−0.85/−0.38]	0.000
	4v0	0.61 * [0.07/1.15]	0.010	−0.11 [−0.35/0.14]	1.000
	4v4	0.07 [−0.27/0.41]	1.000	−0.68 * [−0.84/−0.53]	0.000
	4v4v4	0.92 * [0.54/1.30]	0.000	−0.49 * [−0.66/−0.31]	0.000
	5v0	0.49 * [0.11/0.87]	0.000	−0.18 * [−0.36/−0.01]	0.030
	5v5	0.19 [−0.14/0.52]	1.000	−0.66 * [−0.81/−0.51]	0.000
	5v5v5	1.03 * [0.44/1.62]	0.000	−0.51 * [−0.78/−0.24]	0.000
	Eleven Player Break	−0.03 [−0.58/0.52]	1.000	−0.50 * [−0.76/−0.25]	0.000

Table 5. *Cont.*

		Physiological Load		Biomechanical Load	
		Dif. Mean [I/S]	Sig.	Dif. Mean [I/S]	Sig.
3v3	*3v3v3*	0.77 * [0.32/1.21]	0.000	0.07 [−0.13/0.27]	1.000
	4v0	0.69 * [0.21/1.16]	0.000	0.58 * [0.36/0.79]	0.000
	4v4	0.14 [−0.09/0.38]	1.000	0 [−0.11/0.11]	1.000
	4v4v4	0.99 * [0.71/1.28]	0.000	0.20 * [0.06/0.33]	0.000
	5v0	0.57 * [0.28/0.85]	0.000	0.50 * [0.37/0.63]	0.000
	5v5	0.27 * [0.05/0.48]	0.000	0.02 [−0.08/0.12]	1.000
	5v5v5	1.11 * [0.57/1.64]	0.000	0.17 [−0.07/0.42]	1.000
	Eleven Player Break	0.04 [−0.45/0.53]	1.000	0.18 [−0.05/0.4]	0.620
3v3v3	*4v0*	−0.08 [−0.68/0.51]	1.000	0.51 * [0.23/0.78]	0.000
	4v4	−0.62 * [−1.05/−0.20]	0.000	−0.07 [−0.27/0.12]	1.000
	4v4v4	0.23 [−0.23/0.68]	1.000	0.13 [−0.08/0.34]	1.000
	5v0	−0.20 [−0.66/0.26]	1.000	0.43 * [0.22/0.64]	0.000
	5v5	−0.50 * [−0.91/−0.08]	0.000	−0.05 [−0.24/0.14]	1.000
	5v5v5	0.34 [−0.30/0.98]	1.000	0.1 [−0.19/0.4]	1.000
	Eleven Player Break	−0.73 * [−1.33/−0.12]	0.000	0.11 [−0.17/0.38]	1.000
4v0	*4v4*	−0.54 * [−1.00/−0.08]	0.000	−0.58 * [−0.79/−0.37]	0.000
	4v4v4	0.31 [−0.18/0.80]	1.000	−0.38 * [−0.6/−0.16]	0.000
	5v0	−0.12 [−0.61/0.37]	1.000	−0.08 [−0.3/0.15]	1.000
	5v5	−0.42 [−0.87/0.03]	0.120	−0.56 * [−0.76/−0.35]	0.000
	5v5v5	0.42 [−0.24/1.08]	1.000	−0.40 * [−0.7/−0.1]	0.000
	Eleven Player Break	−0.64 * [−1.27/−0.02]	0.040	−0.40 * [−0.69/−0.11]	0.000
4v4	*4v4v4*	0.85 * [0.59/1.10]	0.000	0.20 * [0.08/0.31]	0.000
	5v0	0.42 * [0.16/0.68]	0.000	0.50 * [0.38/0.62]	0.000
	5v5	0.12 [−0.05/0.29]	1.000	0.02 [−0.06/0.1]	1.000
	5v5v5	0.96 * [0.45/1.48]	0.000	0.18 [−0.06/0.41]	0.980
	Eleven Player Break	−0.10 [−0.57/0.37]	1.000	0.18 [−0.04/0.39]	0.410
4v4v4	*5v0*	−0.43 * [−0.74/−0.12]	0.000	0.31 * [0.16/0.45]	0.000
	5v5	−0.73 * [−0.97/−0.49]	0.000	−0.18 * [−0.28/−0.07]	0.000
	5v5v5	0.11 [−0.43/0.65]	1.000	−0.02 [−0.27/0.23]	1.000
	Eleven Player Break	−0.95 * [−1.45/−0.45]	0.000	−0.02 [−0.25/0.21]	1.000
5v0	*5v5*	−0.30 * [−0.54/−0.06]	0.000	−0.48 * [−0.59/−0.37]	0.000
	5v5v5	0.54 [0.00/1.08]	0.050	−0.33 * [−0.57/−0.08]	0.000
	Eleven Player Break	−0.52 * [−1.03/−0.02]	0.030	−0.32 * [−0.55/−0.09]	0.000
	Partido oficial	−0.66 * [−1.00/−0.32]	0.000	−0.59 * [−0.75/−0.44]	0.000
	Tiros libres	1.54 * [1.16/1.92]	0.000	0.32 * [0.14/0.49]	0.000
5v5	*5v5v5*	0.84 * [0.33/1.34]	0.000	0.15 [−0.08/0.39]	1.000
	Eleven Player Break	−0.23 [−0.69/0.24]	1.000	0.16 [−0.06/0.37]	1.000
5v5v5	*Eleven Player Break*	−1.06 * [−1.73/−0.39]	0.000	0 [−0.3/0.31]	1.000

Notes: The * means statistically significant differences.

4. Discussion

The aim of this article was twofold: firstly, to know and describe the physiological and biomechanical loads of the different tasks used in basketball training and, subsequently, to make a practical proposal of these tasks throughout a typical training week.

In relation to the first goal, the present study has allowed us to categorize the different tasks used in basketball training under the perspective of physiological load or biomechanical load. One of the main reasons for conducting this study is that, as reflected in several specific investigations [2,5], there are still many limitations in the research carried out to date on this topic given the large number of variables that can modify the load imposed by each of the tasks used in basketball. Moreover, this aspect is usually analyzed under the view of high or low load, i.e., under the perspective of the amount of load, but not under the perspective of the nature of the training load, which can be physiological or neuromuscular in nature [23]. For example, in the results of the review by O'Grady et al. (2020) [2], it is pointed out that the results of different studies analyzed [29,44] suggest that SSGs with fewer players (2v2, 3v3) cause a greater training load, both internally and externally, compared to SSGs with a greater number of players (4v4, 5v5), and that exercises used in full court also involve a greater external load than those performed in half court, regardless of team size. Similarly, Clemente (2016) [45] suggests that involving fewer players in SSGs means higher intensity compared to 5v5. Atli et al. (2013) [46] also suggest that when the number of players remains constant but the playing area increases (leading to an increase in the relative distance to be covered), significant differences arise in the load of each of the SSGs. While most of the results found so far are in line with these ideas [29,30,44], they are still very general, because as the results of the present research show, under the perspective of biomechanical and physiological loading, these results can be nuanced, and therefore, would be a better help for coaches when designing training sessions.

Therefore, the results of this study are considered relevant, as it is the first research, to the best of the researchers' knowledge, to classify the different training tasks based on the nature of the load, i.e., physiological load or biomechanical load. The main results obtained can be seen graphically in Figure 1. In summary, it could be said that those tasks that cover more space (full court vs. half court) and with fewer defenders (3vs3, 2vs2, 11 counterattack, 5v0, 4v0, and 3v0) have a higher physiological load, while tasks without defense tend to have lower values of biomechanical load. However, those tasks with less space and more defenders (3v3v3, 4v4v4, 5v5, and 4v4) have a higher biomechanical load.

It could be concluded that the higher biomechanical load is closely related to the presence of defenders. However, in the case of 1v0 and 2v0 tasks, although less demanding, it should be noted that they present a certain biomechanical load (as they are normally linked to technical work and, therefore, accumulate a high number of jumps/min). In this sense, the study by Schelling & Torres. (2016) [29] also found that, for variables such as accelerations per minute, half-court exercises were more demanding. Specifically, 2v2 and 5v5 in half court showed the highest values for accelerations per minute among the different SSGs analyzed. In the study by Olthof et al. (2021) [24], they found a positive association between the biomechanical load of the training sessions with the players' statistics during the match and suggested that biomechanical loads were good predictors for game performance, in the way that excessive biomechanical loads from training may negatively impact game performance. Finally, Castillo et al. (2021) [47] found significant differences in high decelerations and jumps when considering the interaction of the defensive style factors and the outcome of game-based drills.

Although numerous modulators of the external load (opposition/non-opposition, number of opponents, type of opposition, limitation of technical actions, or feedback from the coach) have been described in the specific literature, the playing space seems to be the fundamental variable in the regulation of the intensity of the exercises (i.e., [2,4,45]). The m^2/player ratio determines and guides the task load. By modifying/restricting the absolute (total m^2) or relative (m^2/player) spaces, the biomechanical and physiological demands of the exercises can be modulated to a large extent. However, the results obtained in the present study qualify this idea, as it is not only space that will be the main modulator of the load experienced by the player, but also the presence or absence of defenders. Therefore, the combination of these two variables will be the main modulator of tasks to impose a greater physiological or biomechanical load on the athlete. This could coincide with the results obtained by Sansone et al. (2023) [4], who, while analyzing the training tasks used in basketball from a different perspective, come to the conclusion that the modification of the number of players involved in the task and the space available to the player should be used to modify the external load experienced.

However, in the present study we should avoid having a dichotomous view of this perspective, and it would be much more convenient to understand this analysis not as an analysis of the training tasks between two extremes (high physiological load or high biomechanical load), but as a continuum between these possibilities. In this sense, according to the results obtained, it would be much more advisable to classify the tasks as exercises fundamentally of biomechanical orientation (1v1 full court, 3v3v3, 4v4v4v, and 5v5v5), exercises fundamentally of physiological orientation (3v0 full court, 4v0 full court, and 5v5v5), low-intensity mixed-orientation drills (1v0 half court and 2v0 half court), and high-intensity mixed-orientation drills (official match, 4v4 full court, 3v3 full court, 5v5 full court, counter attack of 11, and 2v2 full court).

In relation to the second goal, it is necessary to highlight its clear practical application, as the present analysis allows us to model training based on the knowledge of the real impact of each task. With this objective we could define different types of sessions as shown in Table 6. Based on the results obtained, depending on the objectives we are looking for when designing the training, we will be able to select more suitable tasks for each of these objectives:

Table 6. Different types of sessions according to the objectives and physical orientation (physiological or biomechanical).

Orientation	Session Duration	Tasks		Task Duration
		Main:	3v0-4v0-5v0	15–20 min
Physiological	60–90 min	Reinforcing:	3v3-4v4-5v5	10–12 min
		Accessories:	1v0-2v0	10–12 min
		Main:	1v1FC-3v3v3-4v4v4-5v5v5	15–20 min
Biomechanical	60–90 min	Reinforcing:	5v5HC	10–12 min
		Accessories:	1v0-2v0	10–12 min
		Main:	2v2 FC-11PB-3v3-4v4-5v5-SGs	15–20 min
Mixed high intensity	60–90 min	Reinforcing:	-	-
		Accessories:	1v0-2v0	10–12 min
		Main:	3v3v3-4v4v4-5v5v5-4v4-5v5	15–20 min

Table 6. *Cont.*

Orientation	Session Duration	Tasks		Task Duration
Tapering I	60–75 min	Reinforcing:	3v3	10–12 min
		Accessories:	1v0-2v0	10–12 min
		Main:	5v5v5-4v4-5v5	15–20 min
Tapering II	45–60 min	Reinforcing:	5v0	10–12 min
		Accessories:	1v0-2v0	10–12 min
		Main:	5v0	8–10 min
Tapering III	30–45 min	Reinforcing:	5v5v5-5v5 (limited contact, no tape)	5–8 min
		Accessories:	1v0-2v0	10–12 min

Another main application of this task classification is the weekly design of training according to the number of competitions and their location, as it appears in Figure 4. Weekly tapering or short-term tapering is the weekly adjustment of the training load with the objective of obtaining an optimal performance for the competition. This programming involves an overload phase (the day's farthest away from the competition) and a tapering phase (the days closest to the competition). Since there is a high sensitivity of physical qualities to tapering in team sports, understanding the differences in the demands of the different tasks allows us to improve the exercise selection system and training design, especially when seeking to optimize weekly performance. It should be noted that there are studies that have revealed a large interindividual variability in individual sports in response to tapering.

Figure 4. Weekly training design according to the number of competitions. Notes: The basketball ball represents an official game according to a typical calendar in basketball leagues.

Greater control of the training stimulus and the adaptations that occur during periods of progressive loading and tapering, especially during periods of intense physiological and psychological stress, that is, prior to competition, could improve the training design and management of the training load. Therefore, the choice of exercises could be crucial to establish an optimal pre-competition physical load.

The main limitation of the present study is that probably not all of the possible tasks to be used in basketball training have been analyzed, although it can be observed that most of the tasks normally used in training sessions are included. However, what is relevant is that it allows for a more correct design of the training sessions by placing the tasks with a greater biomechanical impact at the beginning of the sessions and trying to place the tasks with a greater physiological orientation towards the end of the training. Similarly, if the objective of the planned training is more related to fatigue endurance work, the predominant tasks should be those with a high physiological load, whereas if the objective of the training is mainly tactical or strategic, the tasks to be used will have a high biomechanical load. Additionally, one of the main limitations of this study is the small and highly specific sample size, which restricts the generalizability of the findings. In this regard, the results are closely related to the specific coaching methodology and training design employed in this study, limiting their applicability to other coaching approaches or contexts.

Future research should aim to include larger and more diverse samples to enhance the generalizability of the findings. Expanding the study to include players from various competitive levels, age groups, and geographical regions would provide a better understanding of the physiological and biomechanical demands in basketball training. Furthermore, exploring how different coaching methodologies impact these variables could offer valuable insights for practitioners seeking to adapt the findings to their specific needs and training environments.

5. Conclusions

This study classifies basketball training tasks according to their physiological or biomechanical load, showing that tasks with more space and fewer defenders impose higher physiological loads, while those with less space and more defenders increase the biomechanical load. For training design, it is recommended to place tasks with higher biomechanical load at the beginning of the session and those with physiological orientation toward the end. Manipulating space and the presence of defenders allows for adjusting task intensity to meet specific objectives, optimizing performance and avoiding overtraining.

Author Contributions: Conceptualization, E.A.-P.-C. and A.L.; methodology, E.A.-P.-C. and C.R.; software, C.R. formal analysis, E.A.-P.-C.; investigation, E.A.-P.-C. and C.S.; resources, E.A.-P.-C. and C.S.; data curation, E.A.-P.-C., C.R. and C.S.; writing—original draft preparation, E.A.-P.-C. and C.R.; writing—review and editing, E.A.-P.-C. and A.L.; visualization, E.A.-P.-C. and C.R.; supervision, C.S., X.S. and A.L.; project administration, C.S.; funding acquisition, A.L. All authors have read and agreed to the published version of the manuscript.

Funding: This research received no external funding.

Institutional Review Board Statement: This study was conducted in accordance to the Declaration of Helsinki [32].

Informed Consent Statement: Informed consent was obtained from all subjects involved in the study.

Data Availability Statement: The data used are the property of the basketball team in which they were recorded and are not publicly available. The authors do not have permission to share the data publicly. However, should anyone be interested in learning more, please contact the authors directly.

Conflicts of Interest: The authors declare no conflicts of interest.

References

1. Abdelkrim, B.; El Fazaa, S.; El Ati, J. Time-Motion Analysis and Physiological Data of Elite Under-19-Year-Old Basketball Players During Competition. *Br. J. Sports Med.* **2007**, *41*, 69–75. [CrossRef] [PubMed]
2. O'Grady, C.; Fox, J.; Dalbo, V.; Scanlan, A. A Systematic Review of the External and Internal Workloads Experienced during Games-Based Drills in Basketball Players. *Int. J. Sports Physiol. Perform.* **2020**, *15*, 603–616. [CrossRef] [PubMed]
3. Alonso-Pérez-Chao, E.; Portes, R.; Gómez, M.Á.; Parmar, N.; Lorenzo, A.; Jiménez Sáiz, S.L. A Narrative Review of the Most Demanding Scenarios in Basketball: Current Trends and Future Directions. *J. Hum. Kinet.* **2023**, *89*, 231–245. [CrossRef]
4. Sansone, P.; Gasperi, L.; Makivic, B.; Gomez-Ruano, M.A.; Tessitore, A.; Conte, D. An Ecological Investigation of Average and Peak External Load Intensities of Basketball Skills and Game-Based Training Drills. *Biol. Sport* **2023**, *40*, 649–656. [CrossRef]
5. Russell, J.L.; Mclean, B.D.; Impellizzeri, F.M.; Strack, D.S.; Coutts, A.J. Measuring Physical Demands in Basketball: An Explorative Systematic Review of Practices. *Sports Med.* **2020**, *51*, 81–112. [CrossRef]
6. Alonso, E.; Miranda, N.; Zhang, S.; Sosa, C.; Trapero, J.; Lorenzo, J.; Lorenzo, A. Peak Match Demands in Young Basketball Players: Approach and Applications. *Int. J. Environ. Res. Public Health* **2020**, *17*, 2256. [CrossRef]
7. Portes, R.; Ribas, C.; López García, A.; Alonso Pérez-Chao, E. A Comparison of Training and Competition Physical Demands in Professional Male Basketball Players. *Int. J. Basketb. Stud.* **2023**, *2*, 25–33.
8. Brandão, F.M.; Junior, D.B.R.; da Cunha, V.F.; Meireles, G.B.; Filho, M.G.B. Differences between Training and Game Loads in Young Basketball Players. *Rev. Bras. Cineantropometria E Desempenho Hum.* **2019**, *21*, e59840. [CrossRef]
9. Conte, D.; Kolb, N.; Scalan, A.T.; Santolamazza, F. Monitoring Training Load and Well-Being during the Un-Season Phase in NCAA Division I Men's Basketball. *Int. J. Sports Physiol. Perform.* **2015**, *13*, 1067–1074. [CrossRef]
10. Freitas, C.G.; Aoki, M.S.; Arruda, A.F.S.; Franciscon, C.; Moreira, A. Monitoring Salivary Immunoglobulin A Responses to Official and Simulated Matches in Elite Young Soccer Players. *J. Hum. Kinet.* **2016**, *53*, 107–115. [CrossRef]
11. Lupo, C.; Tessitore, A.; Gasperi, L.; Gomez, M.A.R. Session-RPE for Quantifying the Load of Different Youth Basketball Training Sessions. *Biol. Sport* **2017**, *34*, 11–17. [CrossRef] [PubMed]
12. Scanlan, A.; Dascombe, B.; Kidcaff, A.P.; Peucker, J.; Dalbo, V. Gender-Specific Activity Demands Experienced during Semiprofessional Basketball Game Play. *Int. J. Sports Physiol. Perform.* **2015**, *10*, 618–625. [CrossRef] [PubMed]
13. Torres-Ronda, L.; Ric, A.; Llabres-Torres, I.; De Las Heras, B.; Schelling, X. Position-Dependent Cardiovascular Response and Time-Motion Analysis during Training Drills and Friendly Matches in Elite Male Basketball Players. *J. Stregth Cond. Res.* **2016**, *30*, 60–70. [CrossRef] [PubMed]
14. Abdelkrim, B.N.; Chaouachi, A.; Chamari, K.; Chtara, M.; Castagna, C. Positional Role and Competitive-Level Differences in Elite-Level Men's Basketball Players. *J. Strength Cond. Res.* **2010**, *24*, 1346–1355. [CrossRef]
15. Sosa, C.; Lorenzo, A.; Trapero, J.; Ribas, C.; Alonso, E.; Jimenez, S.L. Specific Absolute Velocity Thresholds during Male Basketball Games Using Local Positional System; Differences between Age Categories. *Appl. Sci.* **2021**, *11*, 4390. [CrossRef]
16. Trapero, J.; Sosa, C.; Shao Liang, Z.; Portes, R.; Gómez Ruano, M.A.; Bonal, J.; Jiménez Saiz, S.; Lorenzo, A. Comparison of the Movement Characteristics Based on Position-Specific Between Semi-Elite and Elite Basketball Players. *Rev. Psicol. Deporte* **2019**, *28*, 140–145.
17. Maggioni, M.A.; Bonato, M.; Stahn, A.; La Torre, A.; Agnello, L.; Vernillo, G.; Castagna, C.; Merati, G. Effects of Ball Drills and Repeated-Sprint-Ability Training in Basketball Players. *Int. J. Sports Physiol. Perform.* **2019**, *14*, 757–764. [CrossRef]
18. Matthew, D.; Delextrat, A. Heart Rate, Blood Lactate Concentration, and Time-Motion Analysis of Female Basketball Players During Competition. *J. Sports Sci.* **2009**, *27*, 813–821. [CrossRef]
19. Edwards, T.; Spiteri, T.; Piggott, B.; Bonhotal, J.; Haff, G.G.; Joyce, C. Monitoring and Managing Fatigue in Basketball. *Sports* **2018**, *6*, 19. [CrossRef]
20. Scanlan, A.; Wen, N.; Tucker, P.; Dalbo, V. The Relationships between Internal and External Training Load Models during Basketball Training. *J. Strength Cond. Res.* **2014**, *28*, 2397–2405. [CrossRef]
21. Coyne, J.O.C.; Gregory Haff, G.; Coutts, A.J.; Newton, R.U.; Nimphius, S. The Current State of Subjective Training Load Monitoring—a Practical Perspective and Call to Action. *Sports Med. Open* **2018**, *4*, 58. [CrossRef] [PubMed]
22. Reina, M.; García-Rubio, J.; Feu, S.; Ibáñez, S.J. Training and Competition Load Monitoring and Analysis of Women's Amateur Basketball by Playing Position: Approach Study. *Front. Psychol.* **2019**, *9*, 2689. [CrossRef]
23. Vanrenterghem, J.; Nedergaard, N.J.; Robinson, M.A.; Drust, B. Training Load Monitoring in Team Sports: A Novel Framework Separating Physiological and Biomechanical Load-Adaptation Pathways. *Sports Med.* **2017**, *47*, 2135–2142. [CrossRef] [PubMed]
24. Olthof, S.B.H.; Tureen, T.; Tran, L.; Brennan, B.; Winograd, B.; Zernicke, R.F. Biomechanical Loads and Their Effects on Player Performance in NCAA D-I Male Basketball Games. *Front. Sports Act. Living* **2021**, *3*, 670018. [CrossRef] [PubMed]
25. Harper, D.J.; McBurnie, A.J.; Santos, T.D.; Eriksrud, O.; Evans, M.; Cohen, D.D.; Rhodes, D.; Carling, C.; Kiely, J. Biomechanical and Neuromuscular Performance Requirements of Horizontal Deceleration: A Review with Implications for Random Intermittent Multi-Directional Sports. *Sports Med.* **2022**, *52*, 2321–2354. [CrossRef]

26. Warathanagasame, P.; Sakulsriprasert, P.; Sinsurin, K.; Richards, J.; McPhee, J.S. Comparison of Hip and Knee Biomechanics during Sidestep Cutting in Male Basketball Athletes with and without Anterior Cruciate Ligament Reconstruction. *J. Hum. Kinet.* **2023**, *88*, 17–27. [CrossRef]
27. Mandorino, M.; Tessitore, A.; Leduc, C.; Persichetti, V.; Morabito, M.; Lacome, M. A New Approach to Quantify Soccer Players' Readiness through Machine Learning Techniques. *Appl. Sci.* **2023**, *13*, 8808. [CrossRef]
28. Mandorino, M.; Tessitore, A.; Lacome, M. Loading or Unloading? This Is the Question! A Multi-Season Study in Professional Football Players. *Sports* **2024**, *12*, 148. [CrossRef]
29. Schelling, X.; Torres, L. Accelerometer Load Profiles for Basketball-Specific Drills in Elite Players. *J. Sports Sci. Med.* **2016**, *15*, 585–591.
30. Svilar, L.; Castellano, J.; Jukic, I. Comparison of 5vs5 Training Games and Match Play Using Microsensor Technology in Elite Basketball. *J. Strength Cond. Res.* **2019**, *33*, 1897–1903. [CrossRef]
31. McKay, A.K.A.; Stellingwerff, T.; Smith, E.S.; Martin, D.T.; Mujika, I.; Goosey-Tolfrey, V.L.; Sheppard, J.; Burke, L.M. Defining Training and Performance Caliber: A Participant Classification Framework. *Int. J. Sports Physiol. Perform.* **2022**, *17*, 317–331. [CrossRef] [PubMed]
32. Harriss, D.J.; Atkinson, G. Ethical Standards in Sport and Exercise Science Research: 2014 Update. *Int. J. Sports Med.* **2014**, *34*, 1025–1029. [CrossRef] [PubMed]
33. Hodder, R.W.; Ball, K.A.; Serpiello, F.R. Criterion Validity of Catapult Clearsky T6 Local Positioning System for Measuring Inter-Unit Distance. *Sensors* **2020**, *20*, 3693. [CrossRef] [PubMed]
34. Luteberget, L.; Spencer, M.; Gilgien, M. Validity of the Catapult ClearSky T6 Local Positioning System for Team Sports Specific Drills, in Indoor Conditions. *Front. Physiol.* **2018**, *9*, 115. [CrossRef] [PubMed]
35. Serpiello, F.R.; Hopkins, W.G.; Barnes, S.; Tavrou, J.; Duthie, G.M.; Aughey, R.J.; Ball, K. Validity of an Ultra-Wideband Local Positioning System to Measure Locomotion in Indoor Sports. *J. Sports Sci.* **2018**, *36*, 1727–1733. [CrossRef]
36. Hoppe, M.W.; Baumgart, C.; Polglaze, T.; Freiwald, J. Validity and Reliability of GPS and LPS for Measuring Distances Covered and Sprint Mechanical Properties in Team Sports. *PLoS ONE* **2018**, *13*, e0192708. [CrossRef]
37. Luteberget, L.; Holme, B.; Spencer, M. Reliability of Wearable Inertial Measurement Units to Measure Physical Activity in Team Handball. *Int. J. Sports Physiol. Perform.* **2017**, *13*, 467–473. [CrossRef]
38. Castellano, J.; Casamichana, D.; Calleja-González, J.; Román, J.S.; Ostojic, S. Reliability and Accuracy of 10 Hz GPS Devices for Short-Distance Exercise. *J. Sports Sci. Med.* **2011**, *10*, 233–234.
39. Johnston, R.J.; Watsford, M.L.; Kelly, S.J.; Pine, M.J.; Spurrs, R.W. Validity and Interunit Reliability of 10 Hz and 15 Hz GPS Units for Assessing Athlete Movement Demands. *J. Strength Cond. Res.* **2014**, *28*, 1649–1655. [CrossRef]
40. Alonso Perez-Chao, E.; Lorenzo, A.; Scanlan, A.; Lisboa, P.; Sosa, C.; Gómez, M.A. Higher Playing Times Accumulated Across Entire Games and Prior to Intense Passages Reduce the Peak Demands Reached by Elite, Junior, Male Basketball Players. *Am. J. Mens Health* **2021**, *15*, 15579883211054353. [CrossRef]
41. Svilar, L.; Castellano, J.; Jukic, I. Load Monitoring System in Top-Level Basketball Team: Relationship between External and Internal Training Load. *Kinesiology* **2018**, *50*, 25–33. [CrossRef]
42. Alonso Pérez-Chao, E.; Gómez, M.Á.; Lisboa, P.; Trapero, J.; Jiménez, S.L.; Lorenzo, A. Fluctuations in Peak Demands across Quarters during Basketball Games. *Front. Psychol.* **2022**, *13*, 868009. [CrossRef] [PubMed]
43. Hopkins, W.; Marshall, S.; Batterham, A.; Hanin, J. Progressive Statistics for Studies in Sports Medicine and Exercise Science. *Med. Sci. Sports Exerc.* **2009**, *41*, 3–12. [CrossRef]
44. Vázquez-Guerrero, J.; Suarez-Arrones, L.; Gómez, D.C.; Rodas, G. Comparing External Total Load, Acceleration and Deceleration Outputs in Elite Basketball Players across Positions during Match Play. *Kinesiology* **2018**, *50*, 228–234. [CrossRef]
45. Clemente, F. Small-Sided and Conditioned Games in Basketball Training. *Strength Cond. J.* **2016**, *38*, 49–58. [CrossRef]
46. Atli, H.; Yusuf, K.; Alemdarog, U.; Fatma, U. A Comparison of Heart Rate Response and Frequencies of Technical Actions Between Half-Court and Full-Court 3—A Side Games in High School Female Basketball Players. *J. Strength Cond. Res.* **2013**, *27*, 352–356. [CrossRef]
47. Castillo, D.; Raya-González, J.; Clemente, F.M.; Conte, D.; Rodríguez-Fernández, A. The Effects of Defensive Style and Final Game Outcome on the External Training Load of Professional Basketball Players. *Biol. Sport* **2021**, *38*, 483–490. [CrossRef]

sensors

MDPI

Article

Analysing Physical Performance Indicators Measured with Electronic Performance Tracking Systems in Men's Beach Volleyball Formative Stages

Joaquín Martín Marzano-Felisatti [1,2,*], Rafael Martínez-Gallego [2,*], José Pino-Ortega [3], Antonio García-de-Alcaraz [4], Jose Ignacio Priego-Quesada [1] and José Francisco Guzmán Luján [2]

[1] Research Group in Sports Biomechanics (GIBD), Department of Physical Education and Sports, Faculty of Physical Activity and Sport Sciences, Universitat de València, 46010 Valencia, Spain; j.ignacio.priego@uv.es
[2] Research Group in Sports Technique and Tactics (GITTE), Department of Physical Education and Sports, Faculty of Physical Activity and Sport Sciences, Universitat de València, 46010 Valencia, Spain; jose.f.guzman@uv.es
[3] Biovetmed & Sportsci Research Group, Department of Physical Activity and Sport, Faculty of Sport Sciences, University of Murcia, 30720 San Javier, Spain; josepinoortega@um.es
[4] SPORT Research Group (CTS-1024), CIBIS (Centro de Investigación para el Bienestar y la Inclusión Social) Research Center, University of Almería, 04120 Almería, Spain; galcaraz@ual.es
* Correspondence: joaquin.marzano@uv.es (J.M.M.-F.); rafael.martinez-gallego@uv.es (R.M.-G.)

Abstract: Sports performance initiation is of significant interest in sports sciences, particularly in beach volleyball (BV), where players usually combine indoor and BV disciplines in the formative stages. This research aimed to apply an electronic performance tracking system to quantify the physical-conditional performance of young male BV players during competition, considering age group (U15 or U19), sport specialisation (indoor or beach) and the set outcome (winner or loser). Thirty-two young male players, categorised by age and sport specialisation, were analysed during 40 matches using electronic performance tracking systems (Wimu PRO™). Data collected were the set duration, total and relative distances covered, and number and maximum values in acceleration and deceleration actions. U19 players and BV specialists, compared to their younger and indoor counterparts, covered more distance (719.25 m/set vs. 597.85 m/set; 719.25 m/set vs. 613.15 m/set) and exhibited higher intensity in terms of maximum values in acceleration (4.09 m/s^2 vs. 3.45 m/s^2; 3.99 m/s^2 vs. 3.65 m/s^2) and deceleration (−5.05 m/s^2 vs. −4.41 m/s^2). More accelerations (557.50 n/set vs. 584.50 n/set) and decelerations (561.50 n/set vs. 589.00 n/set) were found in indoor players. Additionally, no significant differences were found in variables regarding the set outcome. These findings suggest that both age and specialisation play crucial roles in determining a great physical-conditional performance in young players, displaying a higher volume and intensity in external load metrics, whereas indoor players seem to need more accelerations and decelerations in a BV adaptation context. These insights highlight the age development and sport specialisation in young volleyball and BV athletes.

Keywords: sand sports; microtechnology; accelerometer; automatic detection; player load

Citation: Marzano-Felisatti, J.M.; Martínez-Gallego, R.; Pino-Ortega, J.; García-de-Alcaraz, A.; Priego-Quesada, J.I.; Guzmán Luján, J.F. Analysing Physical Performance Indicators Measured with Electronic Performance Tracking Systems in Men's Beach Volleyball Formative Stages. *Sensors* **2024**, *24*, 7524. https://doi.org/10.3390/s24237524

Academic Editors: Bruno Gonçalves, Bruno Emanuel Nogueira Figueira and Diogo Alexandre Martins Coutinho

Received: 14 October 2024
Revised: 20 November 2024
Accepted: 23 November 2024
Published: 25 November 2024

1. Introduction

Beach volleyball (BV) is an outdoor split-court sport played by pairs on the sand, where environmental and contextual variables influence game performance [1,2]. Its popularity has increased recently, prompting sports science researchers to deepen BV analysis from different perspectives [3]. Technical-tactical [4,5], psychological [6], physiological [2], environmental [7] and reglementary [8] aspects of the game have been studied so far. However, physical-conditional parameters have proved to be one of the most studied aspects of the game, from the beginning years [9–11] to the present [12].

Regarding the physical-conditional analysis, BV is an intermittent sport that combines short maximal efforts with prolonged low-intensity recovery periods [2]. These maximum efforts, composed of jumps, defences and displacements with direction changes at high intensity on an irregular and unstable surface such as the sand, make BV a very physically demanding sport [2]. Therefore, players' internal and external loads during competition and training have been studied extensively for performance monitoring, injury prevention, and health control [1,3,13,14].

When discussing physical-conditional aspects of the game, it is necessary to define the concept of load as the intensity, volume and frequency of stimulus experienced by athletes [15,16]. One of BV's most studied external load variables has been the number of jumps [17,18]. Still, new research trends indicate that kinematic analysis of the game could better integrate the external workloads by considering variables such as total distances, number and intensity of accelerations, decelerations or changes in direction by the players during the game [1,3,19,20].

Moreover, physical-conditional variables have been shown to differ in terms of age groups [1,21], game outcomes [3,22] and players' specialisation [3,14,17,21,23]. In terms of age group, a tendency towards longer sets, more jumps and greater distances covered at higher intensities has been found in older categories [1,21]. Concerning game outcomes, winning teams show higher speeds and jumps in blockers and higher adjusting ability, speed, accelerations and decelerations in defenders [3]. As far as players' specialisation is concerned, different approaches can be proposed. Some authors have used the player's team role to differentiate between blockers, defenders and universal players [3,14,17,21,23]. Still, to date, no BV articles have considered the players' specialisation in terms of exclusivity in the BV practice in the formative stages. In this sense, two methodological approaches can be identified: specificity and multidisciplinary [24,25]. Young BV players usually combine indoor and BV seasons in their formative stages. Therefore, two levels of players' specialisation can be determined: players who train indoors and move to BV during the summer season (indoor) or players with a more specialised dedication to BV (beach).

For a better understanding of physical game demands, in the last few years, research studies using electronic performance tracking systems (EPTSs) in BV have emerged [1,3,13,14,16], and this trend can be considered as a starting point in the application of this technology in BV. To the best of the authors' knowledge, these publications have focused on monitoring female BV players, including a case study analysing 99 matches of an elite team [14]; a comparative study (U23 vs. Senior) analysing ten teams at the Australian BV National Championships [1]; two studies focusing on the validation and description of the physical-conditional demands of NCAA players [13,16]; and a study focusing on how contextual variables (player profile, set type, and match outcome) affect data collected with Global Position Systems in six teams during 30 official matches of the Portuguese BV National Championship [3]. Therefore, a gap in the research literature that provides performance indicators obtained with an EPTS in men's BV is recognised, as well as information related to the initial stages.

For all the above reasons, this research aimed to apply EPTS technology to objectively quantify the physical-conditional performance of young male BV players during competition, considering the age group (U15 vs. U19), the players' sport specialisation (indoor vs. beach), and the set outcome (winner vs. loser). The following hypotheses are established: (1) volume and intensity external load variables will increase with the players' age; (2) adjustment variables, such as the number of accelerations and decelerations, will be higher in indoor players; (3) volume and intensity values will be higher in BV specialist players; and (4) intensity external load variables will be higher in the winners' teams.

2. Materials and Methods

2.1. Participants

Thirty-two youth men players were analysed according to their age category (U15 or U19) and sport specialisation (indoor—I, or beach—B): (a) U15-I (n = 8, 12 ± 1 years, 43.7 ± 9.1 kg and 1.55 ± 0.11 m, national league level), (b) U15-B (n = 8, 13 ± 1 years,

57.8 ± 10.7 kg and 1.70 ± 0.10 m, national league level), (c) U19-I (n = 8, 14 ± 1 years, 58.1 ± 6.5 kg and 1.76 ± 0.05 m, national league level), and (d) U19-B (n = 8, 18 ± 1 years, 68.1 ± 8.0 kg and 1.79 ± 0.07 m, national league level). All participants were free from injury during testing and voluntarily participated in the study. Players over 18 years old and legal representatives of minors signed an informed consent form giving their assent to participate. The study was conducted according to the guidelines of the Declaration of Helsinki (2013) and approved by the Ethics Committee of the University of Valencia (ID: 2158717).

2.2. Procedure

Four simulated competitions were organised, one for each group (U15-I, U15-B, U19-I, U19-B), where the 8 players (4 teams) played a total of 10 one-set matches to 21 points, distributed in three phases: round robin, semi-finals and finals (Figure 1). Athletes were briefed on the measurement protocol and the competition system upon arrival. Before each match, the players were instrumented with electronic tracking devices. The starting and finish set time was recorded, and the set outcome was registered, assigning the winning (W) or losing (L) category to the teams based on the game result. A total of 40 four-player matches (n = 160 records) were monitored (20 U15-B records had to be excluded due to technical issues), in which 1310 points (369 in U15-I, 193 in U15-B, 376 in U19-I, 372 in U19-B) and 536 min (166 in U15-I, 74 in U15-B, 133 in U19-I, 163 in U19-B) were played.

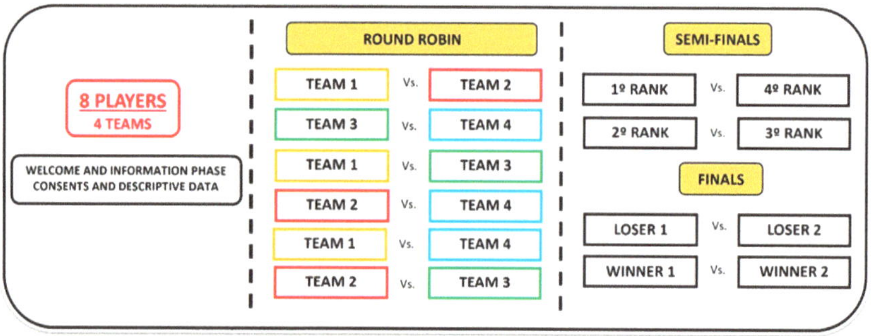

Figure 1. Competition format representation.

2.3. Technology

WIMU PRO^{TM} (RealTrack Systems, Almeria, Spain), a multi-sensor device with four triaxial accelerometers, three triaxial gyroscopes, a triaxial manometer and a Global Position System, was the EPTS used to monitor players' physical-conditional performance during competition [26]. This device was validated previously by different sports scenarios [27,28]. Before each match, players wore a tough-fit top with an interscapular (vertebral T2-T4 level) compartment where the WIMU PRO^{TM} device was located [26]. A push button provided by the manufacturer connected via ANT+ technology to the device was used for more precise signal segmentation (start and end of the set) (Figure 2). Data generated by these sensors during the sets were downloaded and processed in the SPRO program (Software version number: Version 1.0.0 Copilation: 989; RealTrack Systems, Almeria, Spain), where the INTERVAL PRO monitor (Software version number: Version 1.0.0 Copilation: 989; RealTrack Systems, Almeria, Spain) was applied to calculate the variables through algorithms configured by the manufacturer.

Figure 2. Equipment used during competition monitoring.

2.4. Measurement

To compare the players' physical-conditional efforts, nine external load variables were considered to represent the volume and intensity performance: set duration (min), distance covered per set (m/set), distance covered per minute (m/min), number of accelerations and decelerations per set, per minute, and maximum acceleration and deceleration (m/s^2). Variables were selected considering the results of previous studies [1,3,13,14,29].

2.5. Statistical Analysis

A descriptive analysis was carried out in two parts: (i) with the age group (U15 vs. U19), the players' sport specialisation (indoor vs. beach) and the set outcome (winner vs. loser) separately, and (ii) with the interaction between age group and players' sport specialisation (U15-I, U15-B, U19-I, U19-B), and age group and set outcome (U15-L, U15-W, U19-L, U19-W). The median (μ) and interquartile range (IQR) were used for data descriptive representation. The Shapiro–Wilk test showed a non-normal data distribution ($p < 0.05$). Therefore, Mann–Whitney U and Kruskal–Wallis tests were applied ($p < 0.05$), as well as the Holm adjustment for pairwise comparison ($p < 0.05$). Furthermore, the effect size (ES) and the confidence interval (CI) were calculated. Rank-biserial correlation (rbis) effect size considering 0.10 small, 0.30 medium and 0.50 large was used for the Mann–Whitney U test [30]. For the Kruskal–Wallis test, the rank epsilon square (ε^2) effect sizes were chosen with values as follows: 0.00–0.01, negligible; 0.01–0.04, weak; 0.04–0.16, moderate; 0.16–0.36, relatively strong; 0.36–0.64, strong; and 0.64–1.00, very strong [31]. RStudio (version 2023.06.0, package "ggstatplot") software was used in the analysis.

3. Results

The U19 category displayed higher volume and intensity values compared to the U15. The older players covered more distance (719.25 m/set vs. 597.85 m/set, $p = 0.001$, rbis = 0.45) and more relative distance (44.30 m/min vs. 38.66 m/min, $p = 0.001$, rbis = 0.41), and exhibited higher intensity in maximum accelerations (4.09 m/s^2 vs. 3.45 m/s^2, $p = 0.001$, rbis = 0.68) and decelerations (−5.05 m/s^2 vs. −4.41 m/s^2, $p = 0.001$, rbis = 0.40) (Figure 3). Although the variables of duration and the number of accelerations and decelerations per set and minute did not reveal significant differences, there was a trend towards higher values in older age groups. In relation to Figure 4, data show that BV specialists covered a greater distance (719.25 m/set vs. 613.15 m/set, $p = 0.001$, rbis = 0.41) and relative distance (45.59 m/min vs. 37.36 m/min, $p = 0.001$, rbis = 0.69), but performed fewer accelerations (557.50 n/set vs. 584.50 n/set, $p = 0.001$, rbis = 0.34; 34.19 n/min vs. 36.13 n/min, $p = 0.001$, rbis = 0.50) and decelerations (561.50 n/set vs. 589.00 n/set, $p = 0.001$, rbis = 0.34; 34.17 n/min vs. 36.24 n/min, $p = 0.001$, rbis = 0.52) per set and minute. However, they did perform higher maximum accelerations (3.99 m/s^2 vs. 3.65 m/s^2, $p = 0.001$, rbis = 0.31). Moreover, no differences were found considering the set outcome (Figure 5).

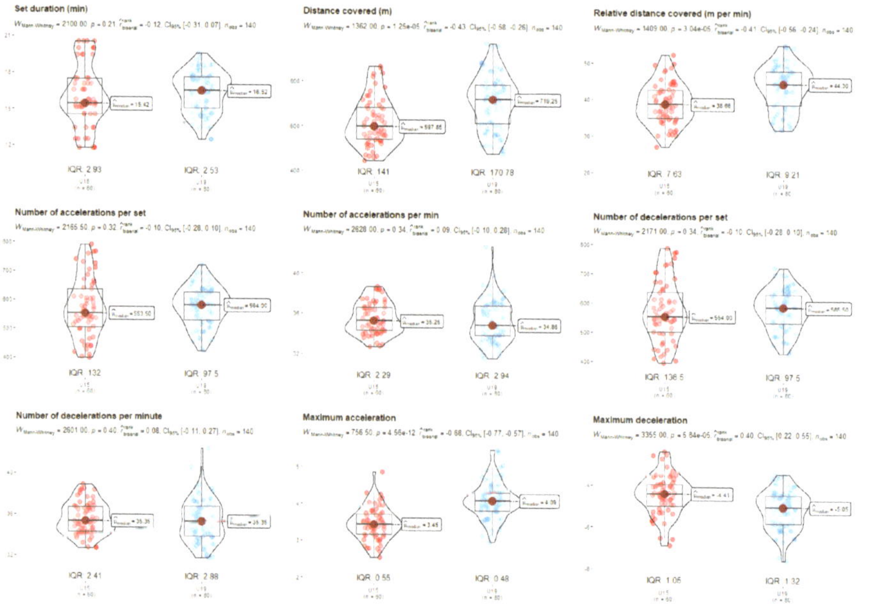

Figure 3. Age group comparison (U15 vs. U19) of the nine performance variables portrayed as violin plots. Median values (μ), interquartile ranges (IQR), Mann–Whitney U test ($p < 0.05$), rank-biserial correlation effect size (rbis), 95% confidence interval ($CI_{95\%}$), and number of observations (n_{obs}) expressed in each plot.

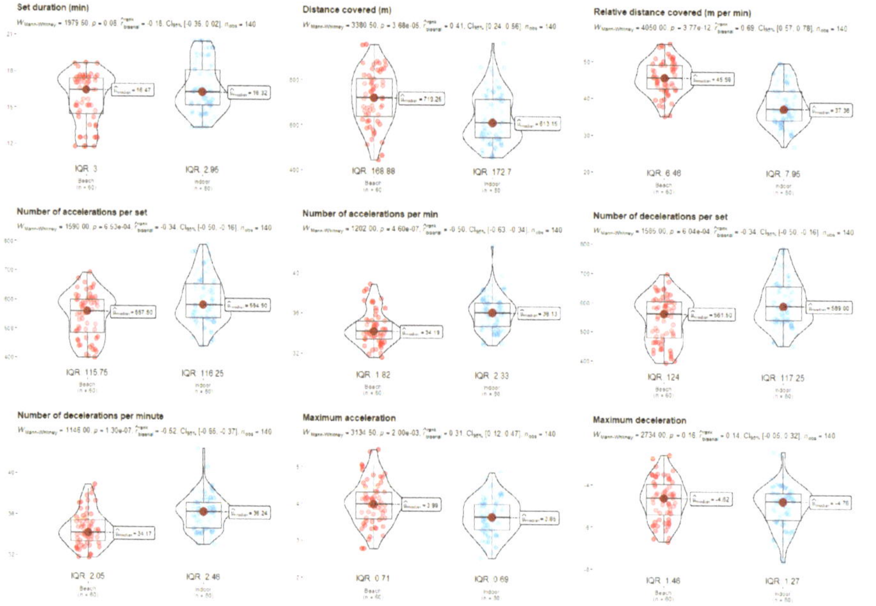

Figure 4. Players' specialisation comparison (beach vs. indoor) of the nine performance variables portrayed as violin plots. Median values (μ), interquartile ranges (IQR), Mann–Whitney U test ($p < 0.05$), rank-biserial correlation effect size (rbis), 95% confidence interval ($CI_{95\%}$), and number of observations (n_{obs}) expressed in each plot.

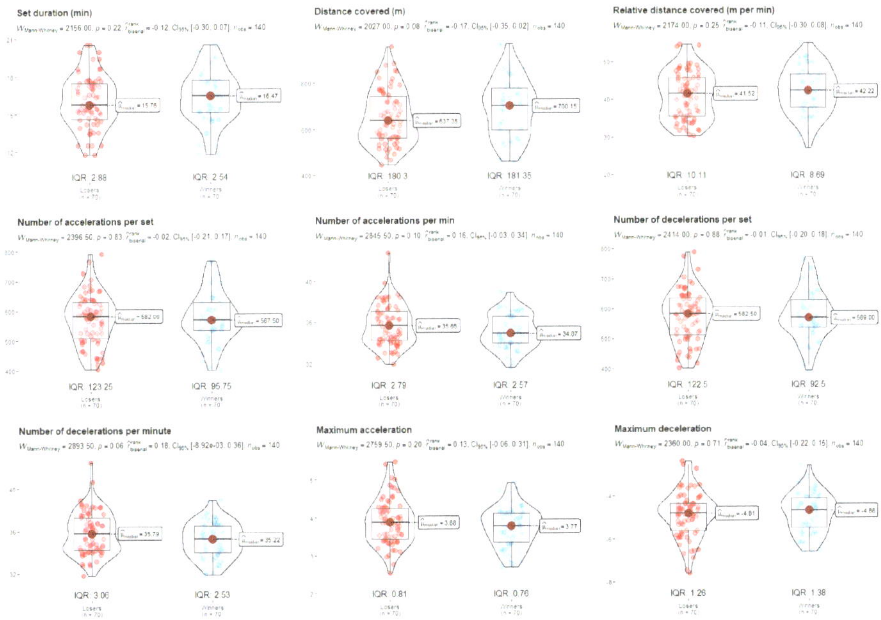

Figure 5. Set outcome comparison (loser vs. winner) of the nine performance variables portrayed as violin plots. Median values (μ), interquartile ranges (IQR), Mann–Whitney U test ($p < 0.05$), rank-biserial correlation effect size (rbis), 95% confidence interval ($CI_{95\%}$), and number of observations (n_{obs}) expressed in each plot.

In the category-level interaction (Table 1), the U19-B group covered the greatest distances (760.90 m/set, $p = 0.001$, $\varepsilon^2 = 0.23$). A tendency to cover more distance with more age and BV specificity was shown. Similar trends were found in relative distance but with higher values in U15-B and U19-B, regardless of age. In terms of accelerations and decelerations (both in volume and intensity), the lower-specialisation groups tended to have higher values. Significant differences were found between the U19-I and the U15-B and U19-B groups, but no differences between U15-I and U19-I were shown. Additionally, U15-I had higher values in decelerations/set compared to U15-B, and U15-I had higher values in the intensity variables (accelerations/min and decelerations/min) compared to U19-B. Maximum values of accelerations and decelerations tended to be higher in older categories, with the higher acceleration values in U19-I.

Table 1. Comparative interaction between age group and players' specialization.

Variables	U15-I μ (IQR)	U15-B μ (IQR)	U19-I μ (IQR)	U19-B μ (IQR)	X^2	p	ε^2	$CI_{95\%}$	Holm
Set duration (min)	15.59 (4.20)	15.12 (4.57)	16.42 (2.79)	16.60 (2.44)	7.68	0.05	0.06	[0.01, 1.00]	†
Distance covered (m/set)	585.95 (129.60)	643.50 (163.45)	682.15 (185.30)	760.90 (163.28)	32.25	0.01	0.23	[0.16, 1.00]	⁂ ‡ #
Relative distance covered (m/min)	35.72 (6.11)	43.95 (7.20)	39.55 (10.31)	46.48 (5.77)	57.71	0.01	0.41	[0.34, 1.00]	* ⁑ ⁂ #
Accelerations per set (n/set)	565.50 (138.75)	506.00 (168.00)	605.00 (84.00)	571.00 (105.25)	14.16	0.01	0.10	[0.04, 1.00]	† #
Accelerations per min (n/min)	35.72 (2.29)	34.69 (1.84)	36.39 (2.06)	33.97 (1.65)	26.99	0.01	0.21	[0.13, 1.00]	⁂ † #

Table 1. *Cont.*

Variables	U15-I μ (IQR)	U15-B μ (IQR)	U19-I μ (IQR)	U19-B μ (IQR)	X^2	p	ε^2	$CI_{95\%}$	Holm
Decelerations per set (n/set)	567.50 (139.25)	498.50 (165.25)	606.50 (88.50)	570.50 (104.75)	14.32	0.01	0.10	[0.05, 1.00]	* † #
Decelerations per min (n/min)	35.94 (2.17)	34.88 (2.03)	36.56 (2.06)	34.13 (2.10)	30.46	0.01	0.22	[0.14, 1.00]	⁂ † #
Maximum acc. (m/s²)	3.45 (0.51)	3.42 (0.49)	3.99 (0.58)	4.16 (0.66)	53.37	0.01	0.38	[0.29, 1.00]	⁑ ⁂ † ‡ #
Maximum dece. (m/s²)	−4.65 (1.23)	−3.79 (0.88)	−5.09 (1.42)	−4.99 (1.28)	26.63	0.01	0.19	[0.14, 1.00]	* † ‡

(U15-I) Under 15 age group indoor specialisation, (U15-B) under 15 age group beach specialisation, (U19-I) under 19 age group indoor specialisation, (U19-B) under 19 age group beach specialisation. (μ) Median values, (IQR) interquartile ranges, (X^2) Kruskal-Wallis ($p < 0.05$), (ε^2) Rank Epsilon Square Effect Sizes, ($CI_{95\%}$) 95% Confidence Interval. Holm adjustment ($p < 0.05$) pairwise differences is represented as: (*) U15-I vs. U15-B, (⁑) U15-I vs. U19-I, (⁂) U15-I vs. U19-B, (†) U15-B vs. U19-I, (‡) U15-B vs. U19-B, (#) U19-I vs. U19-B.

Finally, in the category–result interaction (Table 2), the older age groups' tendency to present higher values in total distance, relative distance, maximum acceleration, and deceleration was confirmed. Moreover, no significant differences were found between the winning and losing teams within each category (U15 vs. U19), confirming the results obtained in Figure 3.

Table 2. Comparative interaction between age group and set outcome.

Variables	U15-I μ (IQR)	U15-B μ (IQR)	U19-I μ (IQR)	U19-B μ (IQR)	X^2	p	ε^2	$CI_{95\%}$	Holm
Set duration (min)	15.42 (2.96)	15.42 (2.69)	16.34 (2.83)	16.34 (2.19)	4.23	0.24	0.03	[0.01, 1.00]	
Distance covered (m/set)	588.60 (115.58)	605.05 (163.76)	687.80 (199.42)	734.70 (109.00)	22.86	0.01	0.16	[0.08, 1.00]	⁑ ⁂ ‡
Relative distance covered (m/min)	37.70 (7.83)	39.23 (8.51)	43.81 (10.52)	44.86 (8.12)	18.74	0.01	0.13	[0.07, 1.00]	⁑ ⁂ ‡
Accelerations per set (n/set)	568.50 (140.75)	541.50 (100.50)	585.00 (114.50)	581.00 (88.75)	1.84	0.61	0.01	[0.01, 1.00]	
Accelerations per min (n/min)	35.73 (2.48)	34.92 (2.52)	35.32 (3.00)	34.84 (3.02)	3.67	0.30	0.03	[0.01, 1.00]	
Decelerations per set (n/set)	572.00 (140.25)	545.00 (104.50)	586.50 (118.25)	585.00 (86.75)	1.72	0.63	0.01	[0.01, 1.00]	
Decelerations per min (n/min)	35.71 (2.42)	35.19 (2.38)	35.79 (3.25)	35.22 (2.69)	4.15	0.25	0.03	[0.01, 1.00]	
Maximum acc. (m/s²)	3.52 (0.42)	3.30 (0.56)	4.14 (0.63)	4.02 (0.42)	50.54	0.01	0.36	[0.28, 1.00]	⁑ ⁂ † ‡
Maximum dece. (m/s²)	−4.38 (0.84)	−4.53 (1.41)	−5.14 (1.26)	−4.95 (1.42)	18.05	0.01	0.13	[0.07, 1.00]	⁑ ⁂ †

(U15-L) Under 15 age group loser team, (U15-W) under 15 age group winner team, (U19-L) under 19 age group loser team, (U19-B) under 19 age group winner team. (μ) Median values, (IQR) Interquartile ranges, (X^2) Kruskal–Wallis ($p < 0.05$), (ε^2) Rank Epsilon Square Effect Sizes, ($CI_{95\%}$) 95% Confidence Interval. Holm adjustment ($p < 0.05$) pairwise differences is represented as: (⁑) U15-L vs. U19-L, (⁂) U15-L vs. U19-W, (†) U15-W vs. U19-L, (‡) U15-W vs. U19-W.

4. Discussion

This study aimed to determine how the physical-conditional variables evolve in men's BV competition, considering age group (U15 vs. U19), players' sport specialisation (indoor vs. beach), and the set outcome (winner vs. loser). It is important to consider that our study aimed to enhance ecological validity by analysing performance variables in real-game contexts and ensuring the findings represent actual competition dynamics. This approach, however, may impact internal validity, as factors like years of training and physical development were not explicitly controlled. While such variables could influence the outcomes, our primary objective was a descriptive performance comparison across levels and categories, prioritising real-world applicability over strict control of confounding factors. The hypotheses proposed by the authors were as follows: (1) volume and intensity of external load variables would increase with players' age, and (3) volume and intensity values would be higher in beach specialist players, both of which were accepted; (2) adjustment variables, such as the number of accelerations and decelerations, would be higher in indoor players, which was partially accepted; and (4) the intensity external load variables would be higher in winning teams, which was rejected.

To start with, older players can sustain higher-intensity efforts for extended periods, achieving greater values in both total and relative distances covered with higher-intensity accelerations and decelerations. These findings align with the existing literature, which has shown that senior players cover greater relative distance, at higher speeds, with more rest periods, and perform more jumps compared to the younger category [1,21]. This suggests an increase in performance metrics as players grow, regardless of the level of specialisation. In this sense, the higher values are attributed to an increase in physical and metabolic capacity as a result of maturation and growth, as well as more experience, shown in better decision-making and skills control [32].

Furthermore, volume and intensity values were higher in BV specialists, as they covered greater distances and relative distances compared to indoor players, as shown in previous studies [12]. Despite performing fewer accelerations and decelerations per set and minute, beach specialist players exhibited higher maximum accelerations. These suggest better specificity adaptation to the BV requirements, playing at a higher intensity without needing extra accelerations and decelerations for adjustments. In this sense, indoor players need to adapt to the open context of the sand surface, wind and sun, as well as to the larger player responsibility area, making it difficult to maintain beach specialist external load volume and intensity [12].

Regarding accelerations and decelerations, indoor players are used to specific position roles, having to adapt in BV to a general role using all volleyball skills (serve, receive, set, spike, block and defend), and needing an additional number of accelerations and decelerations for in-game adjustments [12]. Specifically, the study found that indoor players tend to have slightly higher values in decelerations/set compared to BV specialists in the U15 category and higher accelerations/min and decelerations/min compared to U19 beach specialist players. These findings suggest that while indoor players may perform more frequent adjustments, the differences are not pronounced enough to be statistically significant in all categories.

Finally, no significant differences were found between winning and losing teams regarding the volume and intensity of external load variables within each age category. This finding is consistent across the metrics of total distance, relative distance, maximum acceleration, and deceleration. It provides differing results from recent reference studies on elite senior female athletes, where winning teams show higher speeds and jumping in blockers, and higher adjusting ability, speed, accelerations and decelerations in defenders [3]. This is probably because, in senior elite categories, the physical-conditional aspect of the game becomes more determined than in formative stages. These results suggest that while physical performance is crucial, other factors such as technical skill, tactical execution, and psychological aspects of the game may play more critical roles in determining the outcome of the matches in formative categories.

One major limitation of this study is determining the effect of the number of sets on performance indicators, as the data were collected from one-set simulated competitions. This setup may not accurately reflect the typical accumulative load and fatigue experienced in multi-set matches, although many U15 and U19 competitions present one-set match formats. Moreover, recruiting top elite athletes at this formative stage also becomes a significant challenge, which might have impacted the sample's representativeness.

Moreover, the study was conducted on different days, making it difficult to control contextual variables such as wind, heat, and humidity, which can significantly influence performance. This variability adds a layer of complexity when interpreting the results, as these environmental factors can affect players' physical and technical performances. Furthermore, the study did not control other contextual variables, such as technical, tactical, or psychological aspects, which can also impact performance. These uncontrolled factors may confound the results, making it challenging to isolate the effects of the measured performance indicators. Future research should aim to include a larger, more diverse sample and consider multi-set matches to provide a more comprehensive understanding of performance indicators in men's BV at the formative stages. Additionally, these protocols should be implemented for female formative and professional players to assess potential gender differences, providing valuable insights into performance variations and promoting a more inclusive understanding of the sport. Furthermore, studies should also explore the progression of athletic and technical performance across a broader range of age groups and training levels, as this could clarify how longer training years and competitive exposure influence development in beach volleyball athletes.

5. Conclusions

The findings indicate that age and sports specialisation significantly influence the physical-conditional performance of young volleyball players measured with EPTS technology. Older players and those specialised in BV show higher volume and intensity in their external load metrics, whereas indoor players perform more acceleration and deceleration movements. Additionally, the set outcome does not significantly impact the external load variables.

The findings of this study offer valuable insights for coaches and trainers in developing training programs for young volleyball players. The significant influence of age and sports specialisation on physical-conditional performance suggests that training volume and intensity should be consciously adapted to these factors. Older players and those specialised in BV demonstrate higher volumes and intensities in their physical activities. Therefore, training programs for these athletes should include higher intensity and volume exercises to match their advanced physical capabilities.

Moreover, the finding related to set outcomes highlights the need for coaches to consider other factors, such as technical skills, tactical decisions and psychological aspects of the game, when preparing athletes for competition in the formative stages. Furthermore, by integrating EPTS technology into regular training and match analysis, coaches can establish reference values for different age groups and specialisations, enhancing their ability to monitor and adjust training loads effectively. This approach not only helps optimise performance but also prevents injuries by ensuring that young athletes are not overtraining.

These data and differences found between indoor and BV may be suitable for a proper fusion of experiences in both sports, ensuring a wider physical stimulus favouring future sports specialisation and injury prevention.

Author Contributions: Conceptualization and study design, J.M.M.-F., J.I.P.-Q. and J.F.G.L.; literature review, J.M.M.-F. and A.G.-d.-A.; data collection, J.M.M.-F.; data analysis and interpretation, J.M.M.-F., R.M.-G., J.P.-O. and J.I.P.-Q.; statistical analyses J.M.M.-F., R.M.-G. and J.I.P.-Q.; manuscript writing; J.M.M.-F.; manuscript reviewing, R.M.-G., J.P.-O., A.G.-d.-A., J.I.P.-Q. and J.F.G.L. All authors have read and agreed to the published version of the manuscript.

Funding: The author Joaquín M. Marzano-Felisatti was supported by a pre-doctoral grant from the Ministry of Universities of Spain (Grant number: FPU20/01060).

Institutional Review Board Statement: The study was conducted in accordance with the Declaration of Helsinki and approved by the Ethics Committee of the University of Valencia (protocol code ID: 2158717).

Informed Consent Statement: Informed consent was obtained from all subjects involved in the study.

Data Availability Statement: Data are contained within the article.

Acknowledgments: The authors would like to express their maximum gratitude to the players, coaches, and staff who helped us carry out this project, especially to the Club de Vóley Playa la Malva de Valencia, Club Voleibol Valencia and Club Voleibol Paterna Liceo.

Conflicts of Interest: The authors declare no conflicts of interest.

References

1. Bellinger, P.M.; Newans, T.; Whalen, M.; Minahan, C. Quantifying the Activity Profile of Female Beach Volleyball Tournament Match-Play. *J. Sports Sci. Med.* **2021**, *20*, 142–148. [CrossRef] [PubMed]
2. Magalhaes, J.; Inacio, M.; Oliveira, E.; Ribeiro, J.C.; Ascensao, A.; Magalhães, J.; Inácio, M.; Oliveira, E.; Ribeiro, J.C.; Ascensão, A. Physiological and Neuromuscular Impact of Beach-Volleyball with Reference to Fatigue and Recovery. *J. Sports Med. Phys. Fit.* **2011**, *51*, 66–73.
3. João, P.V.; Medeiros, A.; Ortigao, H.; Lee, M.; Mota, M.P. Global Position Analysis during Official Elite Female Beach Volleyball Competition: A Pilot Study. *Appl. Sci.* **2021**, *11*, 9382. [CrossRef]
4. Giatsis, G.; Schrapf, N.; Koraimann, T.; Tilp, M. Analysis of the Arm Swing Technique during the Spike Attack in Elite Beach Volleyball. *Int. J. Perform. Anal. Sport* **2019**, *19*, 370–380. [CrossRef]
5. Medeiros, A.I.A.; Marcelino, R.; Mesquita, I.M.; Palao, J.M. Performance Differences between Winning and Losing Under-19, under-21 and Senior Teams in Men's Beach Volleyball. *Int. J. Perform. Anal. Sport* **2017**, *17*, 96–108. [CrossRef]
6. Klatt, S.; Rückel, L.-M.; Wagener, S.; Noël, B. Personality Traits and Emotion Regulation Styles of Elite Beach Volleyball Dyads: Examination of Intra-Team Differences, Performance and Satisfaction Levels. *Front. Psychol.* **2021**, *12*, 719572. [CrossRef]
7. Racinais, S.; Alhammoud, M.; Nasir, N.; Bahr, R. Epidemiology and Risk Factors for Heat Illness: 11 Years of Heat Stress Monitoring Programme Data from the FIVB Beach Volleyball World Tour. *Br. J. Sports Med.* **2021**, *55*, 831–835. [CrossRef]
8. Ronglan, L.T.; Grydeland, J. The Effects of Changing the Rules and Reducing the Court Dimension on the Relative Strengths between Game Actions in Top International Beach Volleyball. *Int. J. Perform. Anal. Sport* **2006**, *6*, 1–12. [CrossRef]
9. Batista, G.R.; De Araujo, R.F.; Guerra, R.O. Comparison between Vertical Jumps of High Performance Athletes on the Brazilian Men's Beach Volleyball Team. *J. Sports Med. Phys. Fit.* **2008**, *48*, 172–176.
10. Riggs, M.P.; Sheppard, J.M. The Relative Importance of Strength and Power Qualities to Vertical Jump Height of Elite Beach Volleyball Players during the Counter-Movement and Squat Jump. *J. Hum. Sport Exerc.* **2009**, *4*, 221–236. [CrossRef]
11. Perez-Turpin, J.P.A.; Cortell, J.M.; Chinchilla, J.J.; Cejuela, R.; Suarez, C. Analysis of Jump Patterns in Competition for Elite Male Beach Volleyball Players. *Int. J. Perform. Anal. Sport* **2008**, *8*, 94–101. [CrossRef]
12. Hank, M.; Cabell, L.; Zahalka, F.; Miratsky, P.; Cabrnoch, B.; Mala, L.; Maly, T. Differences in External Load among Indoor and Beach Volleyball Players during Elite Matches. *PeerJ* **2024**, *12*, e16736. [CrossRef] [PubMed]
13. Bozzini, B.N.; McFadden, B.A.; Scruggs, S.K.; Arent, S.M. Evaluation of Performance Characteristics and Internal and External Training Loads in Female Collegiate Beach Volleyball Players. *J. Strength Cond. Res.* **2021**, *35*, 1559–1567. [CrossRef] [PubMed]
14. Nunes, R.F.H.; Carvalho, R.R.; Palermo, L.; Souza, M.P.; Char, M.; Nakamura, F.Y. Match Analysis and Heart Rate of Top-Level Female Beach Volleyball Players during International and National Competitions. *J. Sports Med. Phys. Fit.* **2020**, *60*, 189–197. [CrossRef] [PubMed]
15. Smith, D.J. A Framework for Understanding the Training Process Leading to Elite Performance. *Sports Med.* **2003**, *33*, 1103–1126. [CrossRef]
16. Tometz, M.J.; Jevas, S.A.; Esposito, P.M.; Annaccone, A.R. Validation of Internal and External Load Metrics in NCAA D1 Women's Beach Volleyball. *J. Strength Cond. Res.* **2022**, *36*, 2223–2229. [CrossRef]
17. Natali, S.; Ferioli, D.; La Torre, A.; Bonato, M. Physical and Technical Demands of Elite Beach Volleyball According to Playing Position and Gender. *J. Sports Med. Phys. Fit.* **2019**, *59*, 6–9. [CrossRef]
18. Perez-Turpin, J.A.; Cortell-Tormo, J.M.; Suarez-Llorca, C.; Chinchilla-Mira, J.J.; Cejuela-Anta, R. Relation between Number of Patterns Jumps and Real Time Playing among Elite Female Beach Volleyball Players. *Kinesiol. Slov.* **2009**, *15*, 8–13.
19. Hank, M.; Maly, T.; Zahalka, F.; Dragijsky, M.; Bujnovsky, D. Evaluation of the Horizontal Movement Distance of Elite Female Beach Volleyball Players during an Official Match. *Int. J. Perform Anal. Sport* **2016**, *16*, 1087–1101. [CrossRef]
20. Pelzer, T.; Schmidt, M.; Jaitner, T.; Pfeiffer, M. External Training Load and the Effects on Training Response Following Three Different Training Sessions in Young Elite Beach Volleyball Players. *Int. J. Sports Sci. Coach.* **2020**, *15*, 717–727. [CrossRef]
21. Medeiros, A.; Marcelino, R.; Mesquita, I.; Manuel Palao, J. Physical and Temporal Characteristics of Under 19, Under 21 and Senior Male Beach Volleyball Players. *J. Sports Sci. Med.* **2014**, *13*, 658–665. [PubMed]
22. Da Costa, Y.P.; Paiva-Junior, L.R.; Pereira-Junior, E.F.; Silva, E.L.S.d.; García-de-Alcaraz, A.; Hernández, M.; Batista, G.R. Does the Type of Set and Final Score Change Time Indicators in Beach Volleyball? A Study during Brazilian School Championship. *J. Phys. Educ.* **2022**, *33*, e3318. [CrossRef]
23. Palao, J.; López-Martínez, A.B.; Valades, D.; Ortega, E. Physical Actions and Work-Rest Time in Women's Beach Volleyball. *Int. J. Perform. Anal. Sport* **2015**, *15*, 424–429. [CrossRef]

24. Araújo, D.; Roquette, J.; Davids, K. Ubiquitous Skill Opens Opportunities for Talent and Expertise Development. *Front. Sports Act. Living* **2023**, *5*, 1181752. [CrossRef]
25. Santos, S.D.L.; Memmert, D.; Sampaio, J.; Leite, N. The Spawns of Creative Behavior in Team Sports: A Creativity Developmental Framework. *Front. Psychol.* **2016**, *7*, 1282. [CrossRef]
26. Pino-Ortega, J.; Garcia-Rubio, J.; Ibanez, S.J. Validity and Reliability of the WIMU Inertial Device for the Assessment of the Vertical Jump. *PeerJ* **2018**, *6*, e4709. [CrossRef] [PubMed]
27. Bastida Castillo, A.; Gómez Carmona, C.D.; De la Cruz Sánchez, E.; Pino Ortega, J. Accuracy, Intra- and Inter-Unit Reliability, and Comparison between GPS and UWB-Based Position-Tracking Systems Used for Time-Motion Analyses in Soccer. *Eur. J. Sport Sci.* **2018**, *18*, 450–457. [CrossRef]
28. Pino-Ortega, J.; Reche-Soto, P.; García-de-Alcaraz, A. Comparative Analysis of Recording System for Collecting Kinematic Data in Beach Volleyball. *Proc. Inst. Mech. Eng. Part P J. Sports Eng. Technol.* **2024**, 17543371241229526. [CrossRef]
29. Marzano-Felisatti, J.M.; Pino-Ortega, J.; Priego-Quesada, J.I.; Guzmán-Luján, J.F.; García de Alcaraz, A. Physical Performance and Game Demands in Beach Volleyball: A Systematic Review. *J. Hum. Sport Exerc.* **2025**, *20*, 79–92. [CrossRef]
30. Coolican, H. *Research Methods and Statistics in Psychology*, 5th ed.; Hodder Education Group: London, UK, 2009; ISBN 978-0-340-98344-7.
31. Haque, A.; Ajmeri, N.; Singh, M.P. Understanding Dynamics of Polarization via Multiagent Social Simulation. *AI Soc.* **2023**, *38*, 1373–1389. [CrossRef]
32. Guimarães, E.; Baxter-Jones, A.D.G.; Williams, A.M.; Tavares, F.; Janeira, M.A.; Maia, J. The Role of Growth, Maturation and Sporting Environment on the Development of Performance and Technical and Tactical Skills in Youth Basketball Players: The INEX Study. *J. Sports Sci.* **2021**, *39*, 979–991. [CrossRef] [PubMed]

sensors

MDPI

Article

The Impact of a Congested Match Schedule (Due to the COVID-19 Lockdown) on Creatine Kinase (CK) in Elite Football Players Using GPS Tracking Technology

Jorge Garcia-Unanue [1,*], Antonio Hernandez-Martin [2], David Viejo-Romero [2], Samuel Manzano-Carrasco [3], Javier Sanchez-Sanchez [2], Leonor Gallardo [1] and Jose Luis Felipe [1,4]

1 IGOID Research Group, Physical Activity and Sport Sciences Department, University of Castilla-La Mancha, 45071 Toledo, Spain; leonor.gallardo@uclm.es (L.G.); joseluis.felipe@uclm.es (J.L.F.)
2 School of Sport Sciences, Universidad Europea de Madrid, Villaviciosa de Odón, 28670 Madrid, Spain; antonio.hernandez@universidadeuropea.es (A.H.-M.); david.viejo@universidadeuropea.es (D.V.-R.); javier.sanchez2@universidadeuropea.es (J.S.-S.)
3 Department of Communication and Education, Universidad Loyola Andalucia, 41704 Sevilla, Spain; smanzano@uloyola.es
4 Performance Analysis Department, UD Las Palmas, 35019 Las Palmas de Gran Canaria, Spain
* Correspondence: jorge.garciaunanue@uclm.es

Abstract: The aim was to analyse the consequences of a congested schedule (due to the COVID-19 lockdown) on creatine kinase (CK) in elite football players using GPS tracking technology. A total of 17 elite football players were monitored in training and competition with a global positioning system. Variables including total distance, high-intensity distance, and distance acceleration and deceleration were analysed. Different measurements of serum creatine kinase (CK) concentration were performed on match day (MD) and at 24 h (MD+1), 48 h (MD+2), and 72 h (MD+3) after each match to study the muscle damage of each individual player caused during the match. The results showed a significant increase in physical demands in training (in relative terms regarding the match) at MD+3 compared to MD+1 and MD+2. Furthermore, at +72 h, CK decreases to a value almost the same as that before MD. On the other hand, the players with lower demands for high-intensity actions in the match showed a higher reduction in the concentration of CK at MD+3 compared to MD+1 and MD+2 ($p < 0.05$). It became evident that players with high-intensity demand and a high number of accelerations and decelerations need more time to assimilate the match load and can remain in a state of muscle fatigue for up to 3 days. In addition, a congested schedule can lead to a state of chronic fatigue in elite football players, limiting physical performance and possibly increasing the potential risk of injury for football players.

Keywords: creatine kinase; external load; fatigue; injury risk; physical demands

Citation: Garcia-Unanue, J.; Hernandez-Martin, A.; Viejo-Romero, D.; Manzano-Carrasco, S.; Sanchez-Sanchez, J.; Gallardo, L.; Felipe, J.L. The Impact of a Congested Match Schedule (Due to the COVID-19 Lockdown) on Creatine Kinase (CK) in Elite Football Players Using GPS Tracking Technology. *Sensors* **2024**, *24*, 6917. https://doi.org/10.3390/s24216917

Academic Editors: Bruno Gonçalves, Michael E. Hahn, Bruno Emanuel Nogueira Figueira and Diogo Alexandre Martins Coutinho

Received: 21 June 2024
Revised: 27 September 2024
Accepted: 25 October 2024
Published: 28 October 2024

1. Introduction

Football stands out because of the intermittent physical participation of the players according to their position and the moment of the game, which implies periods in which the players perform high-intensity activity interspersed with actions of lower intensity or recovery [1]. During the 90 min match, each player covers an average total distance of approximately 11 km, with 5% of this distance covered at high speed and 3% at sprint speed [2]. The ability to perform and repeat these intense actions during the match has been considered one of the key factors for the performance of football players [3]. It has also been shown that this characteristic is evolving in football such that high-intensity demands have increased significantly, with an increase in sprint distance of approximately 35% over a period of seven seasons [4]. In addition to the activity carried out at high speed, players perform between 1000 and 1400 actions of high intensity and short duration, ranging between 3 and 5 s, and involve actions with and without the ball that alternate randomly [5]. These efforts require a good number of eccentric muscle contractions, which greatly contribute to the muscular stress suffered by the player and are perceptible up to 120 h after the match [6].

In this sense, monitoring systems have become increasingly important to assess these variables. Among the main monitoring methods, the global positioning system (GPS) is widely used in elite football to quantify training and competition load [7]. GPS is one of the current models of external load control to monitor the movement patterns and physical actions performed by football players during training and matches [8], as well as to help players avoid injuries [9].

The competitive demands of football involve various physiological systems, including the musculoskeletal, nervous, immune, and metabolic systems, to the point that recovery strategies after exercise influence the players' preparation for the next match [10]. Different studies have been carried out with the objective of analyzing this fatigue and recovery, including the analysis-specific variables of performance such as biochemical markers and muscle status, to achieve greater efficiency of progress during recovery [11].

Therefore, these parameters of muscle metabolism, including creatine kinase (CK), lactate dehydrogenase (LDH), and myoglobin, tend to increase after exercise [12]. CK has been used as an indirect marker of muscle damage to quantify and determine the extent of muscle damage caused during competition and training [13]. After a high-intensity effort, the main function of CK in muscle is to provide phosphorus for adenosine triphosphate resynthesis [14]. This parameter increases after exercise, and it peaks between 24 and 48 h post match [15]. Therefore, CK is measured between this period of time after the match when there is no decrease in serum concentration (48 h after), while the next assessment is carried out between 48 and 72 h to determine if the athlete recovers for the next match [16]. An insufficient recovery can adversely affect physical performance [17].

This is an increasingly important aspect in modern football, as it involves a large number of matches during the season, and it is not unusual for a team to play two matches in a week [18]. A congestion of matches can lead to a lack of motivation and concentration, which can affect coordination [19]. Because of low recovery between matches, residual fatigue occurs [20] and increases the stress imposed on the players, which decreases performance [21]. It is necessary to know the impact and physiological changes induced by a football match to help design and develop more effective strategies to shorten the duration to a full recovery [22]. In this regard, further studies are required to demonstrate the impact of high-level match congestion on the muscle damage response of professional players.

Due to the COVID-19 pandemic, there has been a congested fixture period calendar after the resumption of competition in Spain in May 2020. This has influenced the physical performance and injury rate after the quarantine period [23]. Therefore, the quantification of CK and physical demands using GPS devices is ideal to analyse physical performance and muscle damage during the congestion period of training and matches. Therefore, the aim of this study was to analyse the influence of a congested match schedule on physical performance and muscle damage in professional football players after the COVID-19 lockdown.

2. Materials and Methods

2.1. Sample

A longitudinal and quantitative study was performed with male professional football players who played in the Spanish 2nd Division (LaLiga Smartbank) during the 2019–2020 season. All the information was collected, and the data were obtained during the last phase of the LaLiga SmartBank. This period coincides with the resumption of competition after the lockdown period in Spain. Specifically, this period was between 12 June 2020 and 20 July 2020. All the players played 11 matches and participated in 23 training sessions. The study protocol was approved and followed the guidelines established by the local institution, the Ethics Committee of the European University of Madrid (CIPI35/2020), and it was in accordance with the recommendations of the Declaration of Helsinki. The players were previously informed through a document about the purpose of the study and the nature of the tests that would be performed, and an informed consent form was signed prior to the tests.

2.2. Procedures

The data from each player in each match were considered as one observation. Only data of players who participated for at least 10 min in the match and had a complete measurement of physical and physiological variables were included in the study. Furthermore, goalkeepers and players who were penalised or injured were excluded from the analysis and sampling, as were football players not participating in the matches. The players were evaluated individually once before a training session and match competition.

The final study consisted of 17 male professional football players (25.91 ± 3.13 years; 71.27 ± 3.25 kg; 179.36 ± 5.14 cm) divided into three subsamples according to the demand for each of the physical variables analysed (total distance, high-intensity distance [distance travelled above 21 km/h], high-intensity acceleration distance [acceleration distance travelled above 3 m/s^2], and high-intensity deceleration distance [deceleration distance travelled above −3 m/s^2]). The three final subgroups consisted of low-physical-demand players (LPD), medium-physical-demand players (MPD), and high-physical-demand players (HPD).

2.3. Equipment

All team players wore inertial measurement devices (IMUs) (81 × 45 × 16 Mm; 65 gr) with 18 Hz GPS tracking technology (WIMU PRO™, Almería, Spain) to evaluate their movement patterns during each match and training session. For the analysis and data extraction, the software SPROTM v. 960 (REALTRACK SYSTEMS S.L., Almería, Spain) was used. The precision and reliability of this GPS system have been reported in previous investigations [24].

CK measurements were performed at four time points: immediately at the end of the game (MD) as a baseline measure, 24 h after the game (MD+1), 48 h after the game (MD+2), and 72 h after the game (MD+3). Blood samples were collected from the index finger using test strips (REFLOTRON TEST STRIPS®, Roche, Switzerland) and measured on a biochemical analyser (REFLOTRON PLUS®, Roche, Switzerland). The doctor responsible for the medical service was in charge of taking the measurements. Measurements were made according to the established hours, with a variation of ±1 h. Before each sampling, the instrument was calibrated according to the manufacturer's recommendations. A CK measurement protocol was developed according to previous studies [25]. CK sample collection was carried out in the doctor's office. This office was perfectly equipped and prepared, with a standard temperature of 21 °C and a relative humidity of 65%, fulfilling conditions of suitability and homogeneity for carrying out the measurements.

2.4. Statistical Analysis

Results are presented as means ± standard deviations. The three subsamples based on the physical variables (i.e., LPD, MPD, and HPD) were created using k-means clustering. The Kolmogorov–Smirnov test revealed a non-normal behaviour of all variables; therefore, nonparametric tests were performed. The Kruskal–Wallis H test was used to compare the physical parameters and CK blood levels in % relative to the match between MD+1, MD+2, and MD+3. The same test was used to compare the CK blood levels in % relative to the match between MD+1, MD+2, and MD+3 in each of the three subsamples of each physical variable, and the CK blood levels in absolute terms and in % relative to the match between the three subsamples of each physical variable at each of the time points (i.e., MD, MD+1, MD+2, and MD+3). Here, differences were identified, and Dunn–Bonferroni tests were performed for post hoc pairwise comparisons. SPSS V24.0 for Windows (SPSS Inc., Chicago, IL, USA) was used for all tests. The level of significance was set at $p < 0.05$

3. Results

The analysis of variance revealed significant differences in the physical performance and muscle damage of the players in the days after the match compared to the results obtained during the MD (Figure 1; $p < 0.001$). The external load of the players at MD+3 showed a significant increase compared to MD+1 and MD+2 for all the variables analysed ($p < 0.001$; ES: 0.27–0.96). The CK results showed a significant reduction at MD+3 compared to MD+1 (−202.32%; ES: 0.70; $p < 0.001$) and MD+2 (−187.46%; ES: 0.64; $p < 0.001$).

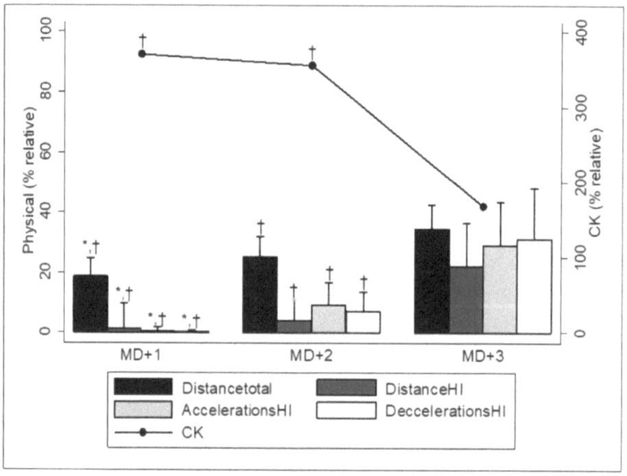

Figure 1. Physical demands and CK blood levels at MD+1 (24 h), MD+2 (48 h), and MD+3 (72 h) according to the MD results. HI: high intensity; CK: creatine kinase. * Significant differences with respect to 48 h; † significant differences with respect to 72 h.

The subgroup analysis revealed a significant influence of the physical demands of the match on the evolution of the CK levels of the players (Table 1; $p < 0.001$). In relation to the total distance covered, the players with the greatest distances covered in the match (HPD) showed a significant reduction at MD+3 compared to MD+1 (−244.37%; ES:0.78) and MD+2 (−191.15%; ES: 0.66) according to the baseline situation ($p < 0.05$). In this sense, in relation to the distance at high intensity and the distance in high-intensity acceleration, the groups with LPD and MPD revealed a reduction in the concentration of CK at MD+3 compared to MD+1 and MD+2 ($p < 0.05$). Finally, the subgroup analysis regarding the distance covered during high-intensity deceleration showed a significant decrease at MD+3 compared to MD+1 in the three groups analysed ($p < 0.05$).

The analysis of the three subgroups only revealed a greater concentration of CK in the MPD group compared to the HPD group (+176.84%; ES: 0.43) as a function of the distance covered during high-intensity acceleration at MD+1 ($p < 0.05$). In absolute terms, the comparative analysis showed higher levels of CK in the blood in the HPD group before the game as a function of the distance travelled at high intensity (+43.66 U l^{-1}; ES: 0.54) and during high-intensity accelerations (+64.24 U l^{-1}; ES: 1.03) compared to the LPD group ($p < 0.05$). In relation to the results obtained 24 h after the match, the MPD group revealed higher blood concentrations of CK than the LPD group (+166.76 U l^{-1}; ES: 1.24) as a function of the distance covered during high-intensity accelerations ($p < 0.05$).

Table 1. CK blood levels at MD, MD+1, MD+2, and MD+3 in professional football players according to the physical demands.

Subgroups		Mean	CK MD	CK MD+1	% MD+1	CK MD+2	% MD+2	CK MD+3	% MD+3
Total distance (m)	LPD	2744.33 ± 1281.18	51.28 ± 59.05	149.80 ± 67.36	619.44 ± 627.35	118.83 ± 124.55	388.24 ± 541.60	17.00 ± 9.90	−56.13 ± 46.40
	MPD	6629.50 ± 1166.34	96.78 ± 77.40	176.80 ± 111.56	179.92 ± 177.63	132.16 ± 103.37	211.49 ± 371.54	63.07 ± 53.39	21.49 ± 199.72
	HPD	9692.32 ± 769.53	89.08 ± 73.70	250.10 ± 214.21	310.34 ± 414.56 ^{c}	207.12 ± 185.08	257.12 ± 371.01 ^{c}	94.73 ± 84.64	65.97 ± 208.53 ^{a,b}
High-intensity distance (m)	LPD	214.54 ± 101.43	74.08 ± 97.18 *	141.71 ± 89.92	355.61 ± 428.16 ^{c}	132.29 ± 126.57	280.31 ± 420.03 ^{c}	59.15 ± 68.52	13.08 ± 182.71 ^{a,b}
	MPD	470.82 ± 74.94	81.63 ± 63.71	261.33 ± 222.65	326.29 ± 396.26 ^{c}	202.50 ± 186.38	291.06 ± 406.10 ^{c}	82.91 ± 71.47	68.57 ± 225.82 ^{a,b}
	HPD	785.53 ± 112.38	117.74 ± 63.48 $^{¥}$	247.49 ± 187.94	251.74 ± 472.24	214.77 ± 175.84	154.78 ± 248.26	126.75 ± 104.24	54.34 ± 157.48
High-intensity acceleration distance (m)	LPD	309.00 ± 102.73	54.39 ± 55.01 *	120.12 ± 61.48 $^{\#}$	358.32 ± 391.37 ^{c}	115.85 ± 92.70	369.37 ± 504.84 ^{c}	50.35 ± 47.87	67.45 ± 275.72 ^{a,b}
	MPD	557.54 ± 59.86	88.48 ± 79.38	286.88 ± 206.80 $^{¥}$	380.93 ± 452.73 ^{c*}	222.04 ± 197.78	263.45 ± 347.54 ^{c}	104.41 ± 96.67	62.92 ± 195.78 ^{a,b}
	HPD	774.93 ± 83.77	118.63 ± 69.24 $^{¥}$	248.30 ± 222.19	204.09 ± 379.59 $^{\#}$	218.47 ± 183.99	150.81 ± 246.22	99.93 ± 67.52	20.67 ± 84.65
High-intensity deceleration distance (m)	LPD	103.55 ± 61.48	69.37 ± 71.06	155.23 ± 75.13	550.61 ± 526.00 ^{c}	152.94 ± 124.41	385.97 ± 462.35 ^{c}	55.00 ± 56.36	−33.09 ± 31.93 ^{a,b}
	MPD	304.47 ± 46.09	87.42 ± 75.03	240.50 ± 227.57	215.20 ± 264.39 ^{c}	185.83 ± 189.00	222.23 ± 356.59 ^{c}	84.38 ± 75.35	65.65 ± 239.01 ^{a,b}
	HPD	443.74 ± 48.30	97.38 ± 72.05	242.64 ± 158.66	426.66 ± 554.48 ^{c}	215.55 ± 156.52	290.23 ± 402.67	103.99 ± 96.34	64.31 ± 151.04 ^{a}

^{a,b,c} Significant differences between MD+1, MD+2, and MD+3, respectively; * Significant differences with HPD group; $^{\#}$ Significant differences with MPD group; $^{¥}$ Significant differences with LPD group; MPD: match day; CK: creatin kinase; LPD: low physical demand players; MPD: medium physical demand players; HPD: high physical demand players.

4. Discussion

The aim of this study was to analyse the influence of a congested match schedule on physical performance and muscle damage in Spanish professional football players after the COVID-19 lockdown. For this fatigue assessment, internal load (serum CK measurements) and external load (GPS variables) were used individually. Our findings demonstrate that players with higher-intensity physical demands in matches, recognized as distance covered at high intensity and distance covered accelerating and decelerating at high intensity, had greater muscle damage and required a longer recovery time. Also, large differences were observed in the recovery time, with 72 h being the ideal time. A long period of inactivity due to the COVID-19 lockdown caused significant negative adaptations in the players, which are related to increases in fitness-dependent injuries [26]. Therefore, a long preseason is of crucial importance because of its protective effect, as it reduces injury risk and injury severity and increases player availability during the season [27]. In this sense, preparing for 4 weeks with limiting guidelines on the return to competition is a handicap to achieve adequate adaptations for competition.

After an analysis of the chronological evolution of the concentration of CK in plasma, an increase in the curve of this biochemical marker is observed from immediately after the end of the match up to 4 days after the end of the match [18]. The highest CK concentration occurred in the 24 h post-match test [28]. In this study, the peak CK levels differ and are lower than reference levels in other studies. This important finding could be due to the players remaining in a state of chronic fatigue for much of this period, which prevented the logical manifestation of this biomarker. Indicators of muscle damage have been shown to decrease as a result of frequent bursts of eccentric loading, or as a result of continued sporting activity [29], as is the case with a congested schedule in football. Therefore, high loads applied over a long period of time by football players can lead to metabolic and organic overloads and, consequently, to chronic fatigue [30].

The release of CK in plasma and its elimination by the body depend on the level of training of the subject and the type, intensity, and duration of exercise [31]. Our results are focused on the characteristics of the effort, which were analysed individually. After 48 h post match, the CK level was still high [15,17]. This suggests that the players did not have a complete recovery 48 h after the match. However, at 72 h, there was a significant reduction in serum CK concentrations compared to previous tests (MD+1 and MD+2). This decrease was evidenced previously [31], where the CK levels decreased in the days after the match, probably as an adaptation to the training stimuli that produced the muscle damage. Therefore, after 72 h, minimal functional recovery and reduced risk of injury can be guaranteed [24]. In this way, physical demands and neuromuscular function show a reduction 24 h and 48 h post match. This decrease in training performance is explained by the structured planning due to the congested match period, and thus the loads are adjusted on the day of the match [20].

On the other hand, a large requirement of the physical demands of the players in the competition influence the concentration of CK and the muscle damage of the players in subsequent days. Changes in the CK levels after the match have higher significant correlations with high-speed running than total distance covered [13]. Our findings indicated a positive relationship between the total distance covered and the time of recovery, where the players with the greatest distance covered in the competition showed a significant reduction at MD+3 compared to groups with less load. This did not happen in other studies that did not find relationships between total distance and the increase in CK concentration in football.18 In addition, other studies have shown a negative correlation between total distance and CK levels at MD+1 and MD+2 [28]. Therefore, players who are stronger and have more strength in the lower body show reduced levels of CK 48 h after the match [32]. This implies that the muscle damage produced by travelling a certain distance does not generate excessive fatigue, which is quantified using concentrations of CK. This may be because players with more fatigue resistance can travel long distances in matches due to the predominance of muscle fibres I or IIb, which are associated with less muscle damage than type II muscle fibres [19].

Otherwise, players who run long distances at high speeds develop increased muscle damage, as demonstrated by post-match CK measurements [17]. These findings are similar to our results, which show significant reductions in the concentrations of CK in the groups with low and medium demand regarding these variables, with respect to the group with HPD in competition. This decrease is especially significant in the MD+3 test, which uses the MD+1 and MD+2 measurements as references. On the other hand, another study showed that high-intensity match activities are related to CK levels at 24 h after the football match [7].

The main take-home application of this study is that it is necessary to adapt the training regimen during the three days after a competition by considering the high load experienced by the players as well as the total volume of minutes played. These players have a high state of fatigue, and their demands during post-match training have been shown to be lower. In addition, it is important that the time saved by the low load of these workouts is used to apply recovery measures. Thus, if there are less than 72 h between competitions, the degree of recovery should be assessed in order to make the lineup and design the match plan. It is possible to save the load, and the performance of these players during the competition is not limited, which takes into account the demands of the previous match.

There are some limitations to be considered when interpreting the present results. Firstly, we do not have data prior to the lockdown to be able to make a pre–post comparison. This is due to the unexpected cancellation of the competition due to COVID-19. Moreover, dietary intake was not monitored and could influence recovery time. Furthermore, factors such as the quality of sleep and the players' activity outside of training hours were also not analyzed.

5. Conclusions

Once the physical results obtained with the GPS devices and the CK values are known, it is essential to manage the workload of players during training with a congested match schedule to ensure adequate recovery for the next match. Special attention should be paid in the days after the match to players who engaged in higher-intensity efforts and high acceleration and deceleration activities during the match. These players will need more time to assimilate the match loads and will not be in an acceptable state of recovery until 3 days after the match. Therefore, with a congested match schedule, a rotation of players is essential for recovery and for the players to be able to perform at an optimal physical level in matches. Organisers of elite tournaments, such as leagues or national team competitions, must be aware of the risks faced by elite players during congested calendar situations and opt for the health and well-being of the players over the spectacle.

Author Contributions: Conceptualization, J.L.F. and J.G.-U.; data curation, J.L.F.; formal analysis, J.G.-U. and A.H.-M.; investigation, S.M.-C., J.L.F. and J.S.-S.; methodology, A.H.-M.; project administration, L.G.; resources, J.S.-S. and J.L.F.; software, J.G.-U. and D.V.-R.; supervision, L.G. and J.L.F.; validation, J.G.-U. and J.L.F.; writing—original draft, S.M.-C.; writing—review and editing, L.G., J.S.-S., J.L.F. and J.G.-U. All authors have read and agreed to the published version of the manuscript.

Funding: A.H.-M. acknowledges the Spanish Ministry of Science, Innovation and Universities for funding the development of his PhD (Grant Number: FPU18/03222). S.M.-C. acknowledges the University of Castilla-La Mancha for funding the development of his PhD (2019/5964). J.G.-U. acknowledges "Fondo Europeo de Desarrollo Regional, Programa Operativo de la Región de Castilla-La Mancha" (2018/11744) for funding the development of his research.

Institutional Review Board Statement: The study protocol was approved and followed the guidelines established by the local institution, the Ethics Committee of the European University of Madrid (CIPI35/2020), and it was in accordance with the recommendations of the Declaration of Helsinki.

Informed Consent Statement: Informed consent was obtained from all subjects involved in the study.

Data Availability Statement: Data that support the findings of this study are available on request from the corresponding author. The data are not publicly available due to contain information that could compromise the privacy of the participants.

Acknowledgments: The authors would like to thank the football sport schools for their support and collaboration, as well as all the football players who contributed to this research.

Conflicts of Interest: The authors declare no conflicts of interest.

References

1. Rebelo, A.; Brito, J.; Maia, J.; Coelho-e-Silva, M.J. Anthropometric characteristics, physical fitness and technical performance of under-19 soccer players by competitive level and field position. *Int. J. Sports Med.* **2013**, *34*, 312–317. [CrossRef] [PubMed]
2. Ehrmann, F.E.; Duncan, C.S.; Sindhusake, D.; Franzsen, W.N.; Greene, D.A. GPS and injury prevention in professional soccer. *J. Strength Cond. Res.* **2016**, *30*, 360–367. [CrossRef] [PubMed]
3. Gabbett, T.; Mulvey, M.J. Time-motion analysis of small-sided training games and competition on elite women soccer players. *J. Strength Cond. Res.* **2008**, *22*, 543–552. [CrossRef] [PubMed]
4. Barnes, C.; Archer, D.T.; Hogg, B.; Bush, M.; Bradley, P. The evolution of physical and technical performance parameters in the English Premier League. *Int. J. Sports Med. Dec.* **2014**, *35*, 1095–1100. [CrossRef]
5. Mohr, M.; Krustrup, P.; Bangsbo, J. Fatigue in soccer: A brief review. *J. Sports Sci.* **2005**, *23*, 593–599. [CrossRef]
6. Nedelec, M.; McCall, A.; Carling, C.; Legall, F.; Berthoin, S.; Dupont, G. The influence of soccer playing actions on the recovery kinetics after a soccer match. *J. Strength Cond. Res.* **2014**, *28*, 1517–1523. [CrossRef]
7. Russell, M.; Sparkes, W.; Northeast, J.; Cook, C.J.; Bracken, R.M.; Kilduff, L.P. Relationships between match activities and peak power output and Creatine Kinase responses to professional reserve team soccer match-play. *Hum. Mov. Sci.* **2016**, *45*, 96–101. [CrossRef]
8. Akenhead, R.; Nassis, G.P. Training load and player monitoring in high-level football: Current practice and perceptions. *Int. J. Sport Physiol.* **2016**, *11*, 587–593. [CrossRef]
9. Ekstrand, J.; Spreco, A.; Windt, J.; Khan, K.M. Are elite soccer teams' preseason training sessions associated with fewer in-season injuries? A 15-year analysis from the union of European football associations (UEFA) elite club injury study. *Am. J. Sports Med.* **2020**, *48*, 723–729. [CrossRef]
10. Reilly, T.; Ekblom, B. The use of recovery methods post-exercise. *J. Sports Sci.* **2005**, *23*, 619–627. [CrossRef]
11. Bishop, P.A.; Jones, E.; Woods, A.K. Recovery from training: A brief review: Brief review. *J. Strength Cond. Res.* **2008**, *22*, 1015–1024. [CrossRef] [PubMed]
12. Warren, J.D.; Blumbergs, P.C.; Thompson, P.D. Rhabdomyolysis: A review. *Muscle Nerve* **2002**, *25*, 332–347. [CrossRef] [PubMed]
13. Young, W.B.; Hepner, J.; Robbins, D.W. Movement demands in Australian rules football as indicators of muscle damage. *J. Strength Cond. Res.* **2012**, *26*, 492–496. [CrossRef] [PubMed]
14. McMahon, S.; Jenkins, D. Factors affecting the rate of phosphocreatine resynthesis following intense exercise. *Sports Med.* **2002**, *32*, 761–784. [CrossRef]
15. Pérez-Castillo Í, M.; Rueda, R.; Bouzamondo, H.; López-Chicharro, J.; Mihic, N. Biomarkers of post-match recovery in semi-professional and professional football (soccer). *Front. Physiol.* **2023**, *14*, 1167449. [CrossRef]
16. Lago-Peñas, C.; Rey, E.; Lago-Ballesteros, J.; Casáis, L.; Domínguez, E. The influence of a congested calendar on physical performance in elite soccer. *J. Strength Cond. Res.* **2011**, *25*, 2111–2117. [CrossRef]
17. Rollo, I.; Impellizzeri, F.M.; Zago, M.; Iaia, F.M. Effects of 1 versus 2 games a week on physical and subjective scores of subelite soccer players. *Int. J. Psychol.* **2014**, *9*, 425–431. [CrossRef]
18. Anderson, L.; Orme, P.; Di Michele, R.; Close, G.L.; Morgans, R.; Drust, B.; Morton, J.P. Quantification of training load during one-, two-and three-game week schedules in professional soccer players from the English Premier League: Implications for carbohydrate periodisation. *Sports Sci.* **2016**, *34*, 1250–1259. [CrossRef]
19. Coutinho, D.; Gonçalves, B.; Wong, D.P.; Travassos, B.; Coutts, A.J.; Sampaio, J. Exploring the effects of mental and muscular fatigue in soccer players' performance. *Hum. Mov. Sci.* **2018**, *58*, 287–296. [CrossRef]
20. Martín-García, A.; Díaz, A.G.; Bradley, P.S.; Morera, F.; Casamichana, D. Quantification of a professional football team's external load using a microcycle structure. *J. Strength Cond. Res.* **2018**, *32*, 3511–3518. [CrossRef]
21. Rey, E.; Lago-Peña, C.; Lago-Ballesteros, J.; Casais, L.; Dellal, A. The effect of a congested fixture period on the activity of elite soccer players. *Biol. Sport* **2010**, *27*, 181–185. [CrossRef]
22. Andersson, H.M.; Raastad, T.; Nilsson, J.; Paulsen, G.; Garthe, I.N.; Kadi, F. Neuromuscular fatigue and recovery in elite female soccer: Effects of active recovery. *Med. Sci. Sports Exer.* **2008**, *40*, 372–380. [CrossRef] [PubMed]
23. Guerrero-Calderón, B. The effect of short-term and long-term coronavirus quarantine on physical performance and injury incidence in high-level soccer. *Soccer Soc.* **2020**, *22*, 85–95. [CrossRef]
24. Gimenez, J.V.; Garcia-Unanue, J.; Navandar, A.; Viejo-Romero, D.; Sanchez-Sanchez, J.; Gallardo, L.; Felipe, J.L. Comparison between two different device models 18 Hz GPS used for time–motion analyses in ecological testing of football. *Int. J. Environ. Res. Public Health* **2020**, *17*, 1912. [CrossRef]

25. Matta, L.; Rhea, M.; Maior, A.S. Physiological evaluation post-match as implications to prevent injury in elite soccer players. *Arch. Med. Deporte: Rev. Fed. Española Med. Deporte Confed. Iberoam. Med. Deporte* **2019**, *36*, 220.
26. Casais-Martínez, L.; Sanroman-Cortes, Z.; Del Hoyo-Lora, M.; Solla-Aguiar, J.; Lago-Penas, C. Return to training and competition after COVID-19 in professional football. *Sci. Rep.* **2020**. Available online: https://sportperfsci.com/wp-content/uploads/2020/06/SPSR103_Lago_final.pdf (accessed on 20 July 2024).
27. Ekstrand, J.; Waldén, M.; Hägglund, M. A congested football calendar and the wellbeing of players: Correlation between match exposure of European footballers before the World Cup 2002 and their injuries and performances during that World Cup. *Br. J. Sports Med.* **2004**, *38*, 493–497. [CrossRef]
28. Freire, L.D.A.; Tannure, M.; Gonçalves, D.; Aedo-Muñoz, E.; Perez, D.I.V.; Brito, C.J.; Miarka, B. Correlation between creatine kinase and match load in soccer: A case report. *J. Phys. Educ. Sport* **2020**, *20*, 1279–1283.
29. Chen, T.; Hsieh, S. The effects of repeated maximal voluntary isokinetic eccentric exercise on recovery from muscle damage. *Res. Q. Exerc. Sport* **2000**, *71*, 260–266. [CrossRef]
30. Radzimiński, Ł.; Jastrzębski, Z.; López-Sánchez, G.F.; Szwarc, A.D.; Dragos, P. Relationships between Training Loads and Selected Blood Parameters in Professional Soccer Players during a 12-Day Sports Camp. *IJERPH* **2020**, *17*, 8580. [CrossRef]
31. Brancaccio, P.; Maffulli, N.; Limongelli, F.M. Creatine kinase monitoring in sport medicine. *Br. Med. Bull.* **2007**, *81*, 209–230. [CrossRef] [PubMed]
32. Owen, A.; Dunlop, G.; Rouissi, M.; Chtara, M.; Paul, D.; Zouhal, H.; Wong, D.P. The relationship between lower-limb strength and match-related muscle damage in elite level professional European soccer players. *J. Sports Sci.* **2015**, *33*, 2100–2105. [CrossRef] [PubMed]

sensors

MDPI

Article

Neuromuscular Load in Professional Women's Handball: Segmentation of the Player Load and the Impacts at Group and Individual Level

Antonio Antúnez [1], Pablo López-Sierra [1,*], Helena Vila-Suárez [2,3] and Sergio J. Ibáñez [1]

1 Grupo de Optimización del Entrenamiento y el Rendimiento Deportivo (GOERD), Universidad de Extremadura, 10003 Cáceres, Spain; antunez@unex.es (A.A.); sibanez@unex.es (S.J.I.)
2 Department of Sports' Special Didactics, Universidade de Vigo, Campus A Xunqueira, S/N, 36005 Pontevedra, Spain; hvila33@gmail.com
3 Grupo de Investigación en Ciencias del Deporte (INCIDE), Departamento de Educación Física y Deportiva, Universidade da Coruña, Oleiros, 15179 A Coruña, Spain
* Correspondence: pablols@unex.es; Tel.: +34-927257827

Abstract: Handball is a team sport characterised by physical interaction with other opponents. This interaction produces a high load on the players that can manifest itself in various ways, from discomfort to prolonged injuries due to tears caused by excessive load. In order to establish correct protocols for application in women's teams, context- and gender-specific reference data must be available. For this reason, the present research aims to find out how women's teams in European competitions prepare for decisive matches during the match week, analysing the load in a segmented way and the level of specificity that should be achieved in training. Ex post facto research was used in which a total of 17 players belonging to a women's first division handball team in Spain participated. The variables player load and impacts extracted from the use of Wimu ProTM inertial devices were analysed. The results showed a high neuromuscular load in players at this competitive level, especially in the variable impacts, reaching values per session of up to 1000 impacts. The individuality analyses show that the load varies significantly depending on the subject, which is why it is considered essential to establish protocols for strength work and load control in the most specific way possible.

Keywords: vertical load; horizontal load; team sport; inertial devices; high performance

Citation: Antúnez, A.; López-Sierra, P.; Vila-Suárez, H.; Ibáñez, S.J. Neuromuscular Load in Professional Women's Handball: Segmentation of the Player Load and the Impacts at Group and Individual Level. *Sensors* **2024**, *24*, 5750. https://doi.org/10.3390/s24175750

Academic Editors: Bruno Emanuel Nogueira Figueira, Diogo Alexandre Martins Coutinho and Bruno Gonçalves

Received: 23 July 2024
Revised: 21 August 2024
Accepted: 23 August 2024
Published: 4 September 2024

1. Introduction

Handball is a high-intensity intermittent sport [1] in which players face during training and competition high efforts that involve a load whose evolution must be controlled and planned by the team's coaching staff. The load planning will be conditioned by the sum of the objective internal load [2], subjective internal load [3], kinematic external load [4] and neuromuscular external load [5].

The neuromuscular load directly affects the performance and health of the players, as this type of load is generated by the execution of practical movements typical of sport in general and handball in particular, such as jumping, running, pivoting, throwing and defending [5]. This load is marked by the dynamic and frequent movements of handball, the numerous impacts that players receive, the numerous changes of direction, the implementation of tactical systems, etc., which can affect injuries caused by overload, such as stress fractures or tendinitis, underlining the importance of the physical-conditional development of players to control muscle fatigue and reduce the risk of injury [6].

The use of inertial devices is well established in various individual and team sports, with handball being one of the least active in the research field [7]. Taking into account that there are articles in handball that have demonstrated the validity of these devices for this sport both in laboratory tests and on the court [8,9], it is essential to investigate the physical load in handball in order to characterise training and competition and optimise the training and performance processes of the players.

The use of these devices and new technologies makes it possible to manage loads efficiently to try to reduce the risk of injury to players [6]. The possibility of collecting multiple variables by obtaining positioning data and accelerometers makes it possible to collect a lot of data with very high precision and frequency, obtaining information on many variables that must subsequently be discriminated. From the use of inertial devices in handball, data have been obtained on player positioning and play distribution [10], throwing speed in running and jumping [11] and even the differences in load in handball on the court and on the beach [12] or the physical load on handball referees [13]. Other studies have pointed out that decelerations could be related to injury risk due to overloading [14].

However, the use of reference values is not sufficient to control the load in team sports. One of the most complex situations for physical trainers in this type of sport is to combine group training with individualised training, as the loads in most situations must be specific to each player due to the heterogeneity between them [15]. Studies in the literature have shown that the load in handball is variable depending on the position and that individual values must be taken into account to improve performance [16] and reduce injury risk [17]. That is why it is necessary to know what types of load should be worked on an individual level and what type of load should be worked on a collective level in order to facilitate and make the training processes more efficient.

Nevertheless, studies in women's handball using new technologies are very scarce and practically nonexistent in professional women's handball. Values obtained in semi-professional women's handball [18] show that players perform in competition with an average of 566.7 total accelerations per game and 419 decelerations. These, weighted to playing time, correspond to 19.06 accelerations per minute and 14.5 decelerations per minute. In order to find values relating to player load and impacts in professional women's handball, it is necessary to consider studies in beach handball [19], where the physical load is far from that of professional handball. In these matches, the average number of impacts is 477.13, being 5.26 higher than 8G. In terms of player load, the values obtained are 14.35 a.u.

Understanding and controlling the neuromuscular load seems essential to manage the loads in professional handball and to achieve the maximum possible performance, as well as to ensure the health of the players. Therefore, the aim of this research was to analyse the objective neuromuscular load in women's handball. In addition, the specific objectives were to find out the segmentation of the loads received by the players during training and to analyse the individuality of the different variables.

2. Materials and Methods

2.1. Design

The present research was classified as an ex post facto design, following the research methods proposed by O'Donoghue [20]. This is because the researchers do not intervene in the training processes, remaining on the sidelines during the session. It is the coaching staff who are in charge of task design and load planning. The researchers simply place the devices before training and remove them at the end of the training session, staying out of the way during the session, only supervising that everything works correctly by visualising the data in real time. The main focus of the research is then taken to the analysis of the data, giving that retrospective naturalisation of the interventions categorised as ex post facto. Due to this, it was a non-experimental research design, taking place in the natural context of sport, without the deliberate manipulation of training or variables during the research process [21].

2.2. Participants

The participants in this research were 17 professional female handball players (age = 25.53 ± 5.69 years; height = 168.35 ± 6.95 cm; weight = 67.88 ± 8.18 kg) belonging to the Spanish top division of women's handball (Liga Guerreras) and the European league (EHF Euro Cup) during the 2022–2023 season. Of the 17 players on the roster, two were goalkeepers, six front lines, four wingers, and five pivots. A non-probabilistic convenience sample was used, as access to these data is very complicated due to the small population of professional athletes. Informed consent was given to all participants before starting the research, explaining the possible risks and benefits of participating in the study. The research was conducted following the criteria of the Declaration of Helsinki (2013) [22], the Ethical Standards in Sport and Exercise Science Research of Harriss et al. (2022) [23] and was approved by the University Bioethics Committee (233/2019). The investigation respected the framework of Organic Law 3/2018 of 5 December on Personal Data Protection and Guarantee of Digital Rights (2018) [24].

Eligibility Criteria

The following criteria were established for the selection of participants: (i) belonging to the team officially, (ii) having participated in at least 80% of the training sessions, and (iii) having been available for at least the last two matches.

Exclusion criteria were: (i) having had a lower body injury less than one month before the start of the measurement, (ii) having trained with lower body discomfort during one or more training sessions, and (iii) having been training in a national team during part of the data collection period.

2.3. Sample

Data were collected from all the training tasks of all the players who met the inclusion and exclusion criteria during one week of a competitive mesocycle in preparation for the regular league and European competition. For the statistical analysis, two databases were created, one for each dependent variable analysed. The total sample analysed is 142 cases collected in a total of 5 training sessions.

2.4. Variables

The independent variable of the study was the training sessions. The dependent variable was the neuromuscular load, specifically the Player Load (measurement based on the accumulation of accelerations in all axes of the plane) and the impacts received by the players measured in G forces. The variable player load corresponds to the vector sum of device accelerations in the 3-axes. The complete formula for its calculation can be found in the article by Reche-Soto et al. [25]. The variables were collected in global values, weighted by time and segmented by work zones. The variables were not manipulated by the researchers during data collection.

2.5. Instruments

The data were collected with Wimu ProTM devices (RealTrack Systems, Almeria, Spain). This required the installation of eight antennas around the field of play, establishing a signal system that allowed triangulating the position and movements of the players during the entire data collection of each session (Figure 1). The devices were positioned with a harness adjusted to the back, at the level of the T2–T4 thoracic vertebrae. These devices are valid and reliable for indoor interventions [26] and were used with 100 hz sampling. The devices included proprietary software for real-time data visualization and proprietary software for retrospective data processing.

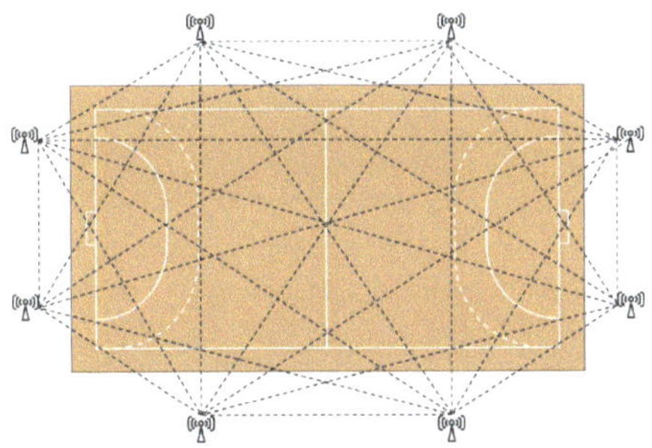

Figure 1. Configuration of the UWB system on the court.

2.6. Procedure

The physical trainers and coaching staff of the club were contacted in order to carry out the research. The players were informed of the benefits and disadvantages that could derive from participation in this research, the study being minimally invasive as the devices did not interfere with the players' sporting practice. When all parties agreed, an informed consent form was drawn up and signed by all participants. This was followed by data collection.

During the training session, the data were monitored in real time through the software included with the inertial devices, *SVivoTM* version 923.4.0 (RealTrack Systems SL, Almeria, Spain, 2020). Once the training was over, the devices were removed and the data were saved in the cloud, stored there for further analysis.

The training sessions were carried out during the first five days of the week, with no rest in between. The sessions had an average duration of 67.4 min, with the longest session being the first one (79 min) and the shortest being the last one (60′). The total distance covered per session averaged 2.64 km per player, again with the longest distance covered being the first session (3.8 km) and the shortest being the last session (1.8 km). On a physical level, the first session was the one that generated the greatest load on the players, since it was the longest, the session in which the greatest distance was covered (and the greatest distance per minute as well), the one that reached the highest values of speed and acceleration, the greatest values of high intensity and player load (total and per minute) and the greatest total maximum impacts. The second session was the one that had more actions with very high G-force values and more intense falls. The third session was the one that generated the greatest imbalances in the right-left footprint. The fourth session was the one with the highest values in high intensity actions per minute and high impacts per minute. The last session was the softest with no maximum values.

Regarding methodological aspects, all sessions included unopposed situations and offensive and defensive Small-Sided Games. The last three also included Full Game situations. The first two sessions were the most balanced in both offensive and defensive aspects, with one unopposed task, two defensive and three offensive Small-Sided Games. The third session was eminently offensive. The fourth session was mainly tactical, with less effective playing time and one task of each type. Finally, the last session was low-load, with attacking and precision situations, and ended with a real situation that lasted half the session.

2.7. Statistical Analysis

A descriptive analysis of both variables was carried out first, in the case of the Player Load segmented by axes and in the case of the impacts segmented by G-force zones. Second, two linear mixed models were performed (one for the Player Load variable and the other for the impacts) to determine the possible differences in the variables, controlling the variance factor between subjects considering the evolution of the week. For this research, the default significance level was set at Alpha (α) = 5%. The level of statistical significance, *p*-value (*p*), was set at $p = 0.05$. Analyses were performed with the statistical programme JAMOVI (v2.3.28, The Jamovi Project, 2022).

3. Results

Table 1 shows the descriptive results of the Player Load variable and its different variables resulting from the segmentation by axes.

Table 1. Descriptive results of the different variables related to the Player Load.

Variable	*N*	*X* ± *SD*	Minimum	Maximum
Player Load (a.u.)	70	48.37 ± 14.29	11.54	78.64
PL/min	70	0.67 ± 0.16	0.22	1.00
Horizontal Player Load (a.u.)	70	31.26 ± 9.21	7.69	49.29
hPL/min	70	0.43 ± 0.11	0.15	0.63
Vertical Player Load (a.u.)	70	31.55 ± 9.53	7.39	53.15
vPL/min	70	0.43 ± 0.11	0.14	0.67
Anteroposterior Player Load (a.u.)	70	19.37 ± 5.65	4.99	30.03
apPL/min	70	0.27 ± 0.06	0.10	0.38

Table 1 shows that the players perform on average 0.67 au/min of Player Load, being mostly vertical and horizontal and in a smaller proportion of anteroposterior component.

Table 2 shows the mixed linear model to analyse the individuality of the variables, including only the total variables without weighting by time.

Table 2. Linear mixed model of Player Load as a function of training session.

Variable	*Marginal* R^2	*Conditional* R^2	*AIC*	*ICC*	*p*
Player Load (a.u.)	0.34	0.85	492.44	0.78	<0.001
Horizontal Player Load (a.u.)	0.34	0.86	428.63	0.78	<0.001
Vertical Player Load (a.u.)	0.33	0.84	438.38	0.77	<0.001
Anteroposterior Player Load (a.u.)	0.37	0.83	367.38	0.72	<0.001

The results of the repeated measures linear mixed model showed a large improvement of the marginal R^2 over the conditional R^2 for all variables, controlling for the random factor of individual responses of subjects with a very high ICC (>0.70).

Table 3 shows the descriptive results of the variable impacts and their values divided by ranges.

Table 3. Descriptive results of the variable impacts.

Variable	*N*	*X* ± *SD*	Minimum	Maximum
Total impacts (counter)	70	7196.91 ± 1757.88	2319	10,440
Total high impacts [>8G] (counter)	70	27.83 ± 19.48	4.00	87
Total impacts [0–2 G]	70	5829.71 ± 1360.27	2031	8347
Total impacts [2–4 G]	70	1002.09 ± 343.82	273	1789
Total impacts [4–6 G]	70	265.51 ± 120.24	10	567
Total impacts [6–8 G]	70	71.77 ± 37.41	1	178
Total impacts [8–10 G]	70	20.23 ± 14.44	3	67
Total impacts [>10 G]	70	7.60 ± 6.14	0	27
Horizontal impacts (counter)	70	7175.43 ± 2212.04	2211	11,308
High horizontal impacts [>8G] (counter)	70	3.50 ± 3.26	0	15
Horizontal impacts [0–2 G]	70	6646.17 ± 1995.46	2127	10,471
Horizontal impacts [2–4 G]	70	464.29 ± 204.71	79	970
Horizontal impacts [4–6 G]	70	52.09 ± 30.36	2	143
Horizontal impacts [6–8 G]	70	9.39 ± 6.76	0	30
Horizontal impacts [8–10 G]	70	2.19 ± 2.23	0	9
Horizontal impacts [>10 G]	70	1.31 ± 1.80	0	11

It can be seen that horizontal impacts account for almost all the impacts received by the handball players. The number of impacts rises to values of over 10,000 impacts per session, with an average of 28 high intensity impacts and a maximum of 27 impacts received by a player in one session.

Table 4 shows the results of the linear mixed model used to analyse the individuality of the players in the different variables.

Table 4. Linear mixed model of impacts as a function of training session.

Variable	*Marginal* R^2	*Conditional* R^2	*AIC*	*ICC*	*p*
Total impacts (counter)	0.38	0.83	1171.38	0.73	<0.001
Total high impacts	0.12	0.68	581.43	0.64	<0.001
Horizontal impacts (counter)	0.35	0.84	1200.57	0.75	<0.001
High horizontal impacts	0.07	0.38	361.16	0.33	<0.146

The results of the repeated measures linear mixed model showed a high improvement in the conditional R^2 versus marginal R^2 for the total impacts and horizontal impacts variables, finding an ICC > 0.70. The high impacts variables also show a significant improvement, especially for high total impacts, with an ICC above 0.05 for high total impacts and < 0.05 for high horizontal impacts.

4. Discussion

The aim of this study was to investigate the neuromuscular load in training handball players through the variables Player Load and Impacts in an elite women's team. From the data, it is clear that the ability to accelerate, decelerate and change direction correctly is necessary for optimal physical performance in handball, which involves intense eccentric contractions that generate neuromuscular fatigue, which in turn leads to identifying the importance of being able to monitor the loads during training for greater control over possible injuries [14].

The descriptive results of the PlayerLoad variable in professional handball players show mean values of 0.67 au/min, reaching a maximum of 1 au/min. Research on handball players that includes the PlayerLoad variable is scarce and has been carried out in competition [27–29]. The studies indicated that Player Load values were similar for wingers, backs and pivots, except in the study by Wik, Luteberget and Spencer [29]. The main results indicate that elite women's handball matches require high physical and physiological demands [28], but by using different devices, it is not possible to establish reference values for intensity and Player Load. Furthermore, different load responses have been recorded across matches, suggesting that coaches should be able to monitor match loads to be able to reproduce them in training in order to optimise the training load prescription according to the demand of each match. Another argument that highlights the importance of monitoring the external load is found in the complexity of controlling loads due to the long duration of the handball season, with the different performance objectives of each team and the possibility of qualifying and playing in international leagues, which conditions the work planning, making it a flexible process over time. To our knowledge, no previous research has focused on workload comparisons during a training week.

In the study presented by Font, Karcher, Reche, Carmona, Tremps and Irurtia [14], using a device similar to the one used in the present research, a season of an elite men's team is monitored, where average Player Load values of 1.1 u.a. in matches throughout the season are shown, higher values than those presented during training by the players in this study. In the article by González-Haro et al. [30], the external load in amateur handball players was analysed, segmenting the load by specific positions. The Player Load values were around 0.59 and 0.83 on average, values that agree with those obtained in the present research. However, the difference between the samples is decisive for the interpretation of these data. The reference values for men's handball are not applicable to women's handball, since the loads borne by female players at the highest competitive level do not correspond to the resulting values for men's handball at the same level [31].

Physical trainers should plan strength and conditioning training by including different types of exercises during training sessions to develop the ability of muscles and tendons to attenuate high eccentric forces, especially in players who tend to play on their backs because of their position and tactical tasks [32]. The assessment of these differences could represent crucial information for handball coaches and team practitioners in order to optimise the training load prescription according to the demand of each league match, both for specific off-court strength training as well as in specific on-court training designed by the coach.

Analysing the Player Load data obtained on the different axes, the similarity between the vertical and horizontal components stands out, as opposed to the differences between the two previous ones with the anteroposterior axis. Previous analyses of the segmentation of player load values according to acceleration axes are not known to exist in the literature. The findings of the present investigation reveal that vertical accelerations are practically the same as horizontal accelerations, so that the load generated by running and track displacements is equivalent to that generated by jumps and vertical oscillations in running. It is important to work on strength and power in all axes of the plane with movements that include free weights and different muscle groups [33], as the use of gym machines that work only one axis in the plane may not be optimal for sports performance in team sports [34].

The horizontal and total impacts received by the players are almost equal, as most of the impacts received by the players come from the interaction with other physical elements in the transverse plane. On average, the players receive more than 1000 impacts of a force between 0 and 4G per training session, but the values remain high until the high impact value (8G), where they are considerably reduced. The values collected in the study by González-Haro, Gómez-Carmona, Bastida-Castillo, Rojas-Valverde, Gómez-López and Pino-Ortega [30] are lower than those presented by the women's team, finding mean values close to 60 impacts/min in the men's amateur team and 80 impacts/min in the women's senior team. These differences are probably due to the competitive level, as the players were training to enter the semi-final of a European competition. Other studies [35] have found differences in the impacts received by female players depending on the specific positions, with these differences being smaller as the hardness of the impacts increased. The intensity of the training sessions, especially with regard to the contact with teammates in the training sessions, influences the neuromuscular load that the players receive, so the loads must be adapted and controlled to the competitive level and players' sex.

Analysis of individual players shows that the neuromuscular load varies greatly between subjects, hence the importance of player-specific load monitoring. The use of subjective tools to control the load for an entire team may not be sufficient in high performance, and other methods must be used to specifically assess the load that each player is able to withstand [36]. Studies such as that of González-Haro, Gómez-Carmona, Bastida-Castillo, Rojas-Valverde, Gómez-López and Pino-Ortega [30] found significant differences in the load depending on the specific positions, which could be a start to individualise in small work groups, but always trying to seek the greatest possible individualisation with the necessary technological support to do so.

Having objective information on the kinematic and neuromuscular external load demands of professional handball players will allow coaches at this level to design training programmes tailored to the specific demands of handball and its tactical game characteristics. These benchmark values are the first of their kind for this population of high-level professional players and should help to adjust training processes to match the demands of competition. However, PlayerLoad data should be interpreted and used with caution, and each brand uses a different algorithm to calculate this variable (some manufacturers calculate PlayerLoad from three-dimensional accelerometer data, while others use two-dimensional data for the calculation) [37].

The main limitation of this research is the transversality of the measurement, as a top-level team is measured in the most demanding training period of the season (preparation for a knockout semi-final match in European competition). However, this limitation is compensated for by the high quality of the data obtained, as it is a very representative week in terms of training and workload. As a prospective for the future, it is recommended that researchers carry out longer measurements and monitor competitions whenever possible.

5. Conclusions

Handball is a sport in which players bear high neuromuscular loads, essentially derived from the accumulation of accelerations in the horizontal and vertical planes and impacts in the horizontal axis. Strength work and training tasks should include complete movements in multiple planes, as well as work with free weights in the gym, combined with actions that involve a physical relationship with the environment.

6. Practical Applications

The results of this study underline the importance of monitoring training sessions to know the fatigue derived from accelerations and impacts in handball players, since they have a great importance when generating load in the players. This load should be taken into account to establish recovery and injury prevention plans during the season.

Due to the evidence of the high neuromuscular loads that players endure, it is recommended to perform off-court strength and power work based on multiaxis work with free weights and combining different muscle groups, reducing the work on a single plane of the gym machines. This off-court work should always be complemented with on-court work with tasks with many players in which there is continuous physical contact with teammates. Research should now focus on generating reference data in competition adapted to each level and sport context, paying special attention to impacts in the horizontal plane, in order to prepare players to withstand the eccentric load derived from the repetition of high G-force impacts throughout training or competition.

Finally, in the literature, consensus should be established on how to reference the load values derived from the Player Load with different devices, as the scarce evidence in this sport regarding this variable makes it difficult to apply reference values when using devices of different brands.

Author Contributions: Conceptualization, A.A., P.L.-S., H.V.-S. and S.J.I.; methodology, P.L.-S. and S.J.I.; software, S.J.I.; validation, A.A. and S.J.I.; formal analysis, P.L.-S.; investigation, P.L.-S. and H.V.-S.; resources, A.A., H.V.-S. and S.J.I.; data curation, P.L.-S. and S.J.I.; writing—original draft preparation, P.L.-S. and H.V.-S.; writing—review and editing, A.A. and S.J.I.; visualization, A.A., P.L.-S., H.V.-S. and S.J.I.; supervision, A.A. and S.J.I.; project administration, A.A. and S.J.I.; funding acquisition, P.L.-S. and S.J.I. All authors have read and agreed to the published version of the manuscript.

Funding: This study has been partially funded by the Aid to Research Groups (GR21149) of the Regional Government of Extremadura (Regional Ministry of Employment and Infrastructures); with the contribution of the European Union through the European Regional Development Funds (ERDF). The author Pablo López-Sierra is a grantee of the "Plan Propio de Iniciación a la Investigación, Desarrollo Tecnológico e Innovación" of the University of Extremadura.

Institutional Review Board Statement: This investigation was approved by the Extremadura University Bioethics Committee (233/2019).

Informed Consent Statement: Informed consent was obtained from all subjects involved in the study.

Data Availability Statement: Data are unavailable due to ethical restrictions.

Conflicts of Interest: The authors declare no conflicts of interest.

References

1. Park, J.; Yoon, K.; Kim, J.-E. Movement Analysis of Women's Handball Players by Position using Inertial Measurement Units. *J. Korea Converg. Soc.* **2020**, *11*, 343–350. [CrossRef]
2. Ortega-Becerra, M.A.; Asián-Clemente, J.A.; López-Adarve, C. The use of TRIMPS to quantify training load in small-sided games in handball. *E-Balonmano Com.* **2016**, *12*, 53–64.
3. Pedersen, A.; Randers, M.B.; Luteberget, L.S.; Moller, M. Validity of Session Rating of Perceived Exertion for Measuring Training Load in Youth Team Handball Players. *J. Strength Cond. Res.* **2023**, *37*, 174–180. [CrossRef]
4. Cobos, D.L.; Ortega-Becerra, M.; Daza, G.; Sánchez-Sáez, J.A. Internal and External Load in International Women's Beach Handball. Official and Unofficial Competition. *Apunt. Educ. Fis. Deportes* **2023**, *151*, 79–87. [CrossRef]
5. Ronglan, L.T.; Raastad, T.; Borgesen, A. Neuromuscular fatigue and recovery in elite female handball players. *Scand. J. Med. Sci. Sports* **2006**, *16*, 267–273. [CrossRef] [PubMed]

6. Nevot-Casas, O.; Pujol-Marzo, M.; Moreno-Planes, B.; Fort-Vanmeerhaeghe, A. Neuromuscular fatigue effects on Hamstring to Quadriceps Ratio in young female players. *Apunt. Med. L'esport* **2020**, *55*, 137–142. [CrossRef]
7. Camomilla, V.; Bergamini, E.; Fantozzi, S.; Vannozzi, G. Trends Supporting the In-Field Use of Wearable Inertial Sensors for Sport Performance Evaluation: A Systematic Review. *Sensors* **2018**, *18*, 873. [CrossRef] [PubMed]
8. Luteberget, L.S.; Holme, B.R.; Spencer, M. Reliability of Wearable Inertial Measurement Units to Measure Physical Activity in Team Handball. *Int. J. Sports Physiol. Perform.* **2018**, *13*, 467–473. [CrossRef]
9. Soler-López, A.; García-de-Alcaraz, A.; Moreno-Villanueva, A.; Pino-Ortega, J. Concurrent Validity and Reliability of Devices to Measure Jump Height in Men's Handball Players. *Sensors* **2022**, *22*, 9070. [CrossRef]
10. Lefevre, T.; Guignard, B.; Karcher, C.; Reche, X.; Font, R.; Komar, J.; Musa, R.M. A deep dive into the use of local positioning system in professional handball: Automatic detection of players' orientation, position and game phases to analyse specific physical demands. *PLoS ONE* **2023**, *18*, e0289752. [CrossRef]
11. Skejo, S.D.; Bencke, J.; Moller, M.; Sorensen, H. Estimating Throwing Speed in Handball Using a Wearable Device. *Sensors* **2020**, *20*, 4925. [CrossRef] [PubMed]
12. Mancha-Triguero, D.; González-Espinosa, S.; Córdoba, L.G.; García-Rubio, J.; Feu, S. Differences in the physical demands between handball and beach handball players. *Rev. Bras. Cineantropometria Desempenho Hum.* **2020**, *22*, e72114. [CrossRef]
13. Gutierrez-Vargas, R.; Ugalde-Ramírez, J.A.; Rojas-Valverde, D.; Müller-Thyssen, M.; Pino-Ortega, J. External and Internal Load of Costa Rican Handball Referees According to Sex and Game Periods. *E-Balonmano Com.* **2021**, *17*, 153–162. [CrossRef]
14. Font, R.; Karcher, C.; Reche, X.; Carmona, G.; Tremps, V.; Irurtia, A. Monitoring external load in elite male handball players depending on playing positions. *Biol. Sport* **2021**, *38*, 475–481. [CrossRef]
15. Karcher, C.; Buchheit, M. On-Court Demands of Elite Handball, with Special Reference to Playing Positions. *Sports Med.* **2014**, *44*, 797–814. [CrossRef] [PubMed]
16. Luteberget, L.S.; Trollerud, H.P.; Spencer, M. Physical demands of game-based training drills in women's team handball. *J. Sports Sci.* **2018**, *36*, 592–598. [CrossRef] [PubMed]
17. Mónaco, M.; Gutiérrez, J.A.; Montoro, J.B.; Til, L.; Drobnic, F.; Nardi, J.; Puigdellivol, J.; Pedret, C.; Rodas, G. Epidemiología lesional del balonmano de elite: Estudio retrospectivo en equipos profesional y formativo de un mismo club. *Apunt. Med. L'esport* **2014**, *49*, 11–19. [CrossRef]
18. García-Sánchez, C.; Navarro, R.M.; de la Rubia, A.; Tannion, K. Influence of playing position on accelerations and decelerations performed by semi-professional female handball players during official matches. *E-Balonmano Com.* **2024**, *20*, 127–134. [CrossRef]
19. Gómez-Carmona, C.D.; García-Santos, D.; Mancha-Triguero, D.; Antúnez, A.; Ibáñez, S.J. Analysis of sex-related differences in external load demands on beach handball. *Rev. Bras. Cineantropometría Desempenho Hum.* **2020**, *22*, e71448. [CrossRef]
20. O'Donoghue, P. *Research Methods for Sports Performance Analysis*; Routledge: New York, NY, USA, 2010.
21. Montero, I.; León, O.G. A guide for naming research studies in Psychology. *Int. J. Clin. Health Psychol.* **2007**, *7*, 847–862.
22. Association, W.M. Declaration of Helsinki: Ethical principles for medical research involving human subjects. *JAMA* **2013**, *310*, 2191–2194. [CrossRef]
23. Harriss, D.J.; Jones, C.; MacSween, A. Ethical Standards in Sport and Exercise Science Research: 2022 Update. *Int. J. Sports Med.* **2022**, *43*, 1065–1070. [CrossRef] [PubMed]
24. Agencia Estatal Boletin Oficial del Estado. *Ley Orgánica 3/2018, de 5 de Diciembre, de Protección de Datos Personales y Garantía de los Derechos Digitales*; Ley Orgánica 3/2018; BOE: Madrid, Spain, 2018. (In Spanish)
25. Reche-Soto, P.; Cardona-Nieto, D.; Díaz-Suárez, A.; Bastida-Castillo, A.; Gómez-Carmona, C.; García-Rubio, J.; Pino-Ortega, J. Player Load and Metabolic Power Dynamics as Load Quantifiers in Soccer. *J. Hum. Kinet.* **2019**, *69*, 259–269. [CrossRef] [PubMed]
26. Gómez-Carmona, C.; Bastida-Castillo, A.; González-Custodio, A.; Olcina, G.; Pino-Ortega, J. Using an Inertial Device (WIMU PRO) to Quantify Neuromuscular Load in Running: Reliability, Convergent Validity, and Influence of Type of Surface and Device Location. *J. Strength Cond. Res.* **2020**, *34*, 365–373. [CrossRef] [PubMed]
27. Luteberget, L.S.; Spencer, M. High-Intensity Events in International Women's Team Handball Matches. *Int. J. Sports Physiol. Perform.* **2017**, *12*, 56–61. [CrossRef] [PubMed]
28. Kniubaite, A.; Skarbalius, A.; Clemente, F.M.; Conte, D. Quantification of external and internal match loads in elite female team handball. *Biol. Sport* **2019**, *36*, 311–316. [CrossRef]
29. Wik, E.H.; Luteberget, L.S.; Spencer, M. Activity Profiles in International Women's Team Handball Using PlayerLoad. *Int. J. Sports Physiol. Perform.* **2017**, *12*, 934–942. [CrossRef]
30. González-Haro, P.J.; Gómez-Carmona, C.D.; Bastida-Castillo, A.; Rojas-Valverde, D.; Gómez-López, M.; Pino-Ortega, J. Analysis of playing position and match statusrelated differences in external load demands on amateur handball: A case study. *Rev. Bras. Cineantropometria Desempenho Hum.* **2020**, *22*, e71427. [CrossRef]
31. Michalsik, L.B.; Aagaard, P. Physical demands in elite team handball: Comparisons between male and female players. *J. Sports Med. Phys. Fit.* **2015**, *55*, 878–891.
32. Lijewski, M.; Burdukiewicz, A.; Pietraszewska, J.; Stachon, A.; Andrzejewska, J.; Chromik, K. Anthropometric and strength profiles of professional handball players in relation to their playing position—multivariate analysis. *Acta Bioeng. Biomech.* **2019**, *21*, 147–155. [CrossRef]
33. Wagner, H.; Gierlinger, M.; Adzamija, N.; Ajayi, S.; Bacharach, D.W.; von Duvillard, S.P. Specific Physical Training in Elite Male Team Handball. *J. Strength Cond. Res.* **2017**, *31*, 3083–3093. [CrossRef] [PubMed]

34. Vojinovic, A.; Janicijevic, D.; Petrovic, M.R.; Garcia-Ramos, A.; Simic, M.; Suzovic, D. Free Weight Training vs. Elastic Band Training: What is a more Effective Strategy for Increasing Maximal Velocity Ability During Handball Throws? *Kinesiology* **2023**, *55*, 21–29. [CrossRef]
35. Manchado, C.; Tortosa-Martínez, J.; Marcos-Jorquera, D.; Gilart-Iglesias, V.; Pueo, B.; Chirosa-Rios, L.J. Monitoring External Load During Real Competition in Male Handball Players Through Big Data Analytics: Differences by Playing Positions. *Kinesiology* **2024**, *56*, 12–23. [CrossRef]
36. Henze, A.S.; Mauch, F. Monitoring and load management in team handball. *Sportverletz. Sportschaden* **2020**, *34*, 136–146. [CrossRef] [PubMed]
37. Torres-Ronda, L.; Beanland, E.; Whitehead, S.; Sweeting, A.; Clubb, J. Tracking Systems in Team Sports: A Narrative Review of Applications of the Data and Sport Specific Analysis. *Sports Med. Open* **2022**, *8*, 15. [CrossRef]

sensors

MDPI

Article

Assessing Variations in Positional Performance across Age Groups and during Matches in Youth Association Football Competitions

Quanchen Liu [1], Zhuhang Huang [1,*], Diogo Coutinho [2,3], Xiaobin Wei [4], Tao An [5] and Bruno Gonçalves [6,7,8]

1 China Football College, Beijing Sport University, Beijing 100084, China; quanchen1993@outlook.com
2 Department of Sports Sciences and Physical Education, University of Maia, 4475-690 Maia, Portugal; dcoutinho@umaia.pt
3 Research Center in Sports Sciences, Health Sciences and Human Development, CIDESD, 5000-801 Vila Real, Portugal
4 School of Strength and Conditioning Training, Beijing Sport University, Beijing 100084, China; weixiaobin@bsu.edu.cn
5 Sport Development Center of Tengzhou City, Tengzhou 277500, China; tztyzx@163.com
6 Departamento de Desporto e Saúde, Escola de Saúde e Desenvolvimento Humano, Universidade de Évora, 7004-516 Évora, Portugal; bgoncalves@uevora.pt
7 Comprehensive Health Research Centre (CHRC), Universidade de Évora, 7004-516 Évora, Portugal
8 Portugal Football School, Portuguese Football Federation, 1495-433 Oeiras, Portugal
* Correspondence: bsuhzh@126.com

Abstract: This study aimed to explore how positional performance varies across different youth age groups and during matches in football competitions. The study encompassed 160 male outfield youth football players ($n = 80$, under-13, U13; $n = 80$, under-15, U15) who belonged to the starting line-up and played the entire first half of each match. The players' positional data were gathered through the global positional system for each of the eight matches performed by each age group. The frequency of near-in-phase synchronization based on speed displacements, spatial exploration index, and the distance to the nearest teammate and opponent were used as variables. Additionally, each match half was segmented into three equal parts to assess changes over time and used as a period factor along with age group. The results indicated that U13 players showed a significant decrease (from small to large ES) in synchronization speed and spatial exploration index throughout the first half of the match, along with a decrease in the distance to the nearest opponent. In contrast, U15 players exhibited most changes during the third segment of the half, with a decrease in speed synchronization and spatial exploration, but an increase in the distance and regularity to the nearest opponent. Comparing both age groups revealed significant differences in speed synchronization across the entire half of the match and within each segmented period (from small to large ES), with U13 consistently showing higher values. The study highlights that long durations in 11 vs. 11 matches might not provide an appropriate learning environment in the U13 age group. Conversely, the U15 group displayed better capacity for tactical adjustments over time, suggesting a higher level of tactical maturity. Overall, these findings emphasize the importance of adapting youth football training and competition structures to the developmental needs and capabilities of different age groups to optimize learning and performance outcomes.

Keywords: global position system; collective behaviour; tactical analysis; youth players

Citation: Liu, Q.; Huang, Z.; Coutinho, D.; Wei, X.; An, T.; Gonçalves, B. Assessing Variations in Positional Performance across Age Groups and during Matches in Youth Association Football Competitions. *Sensors* **2024**, *24*, 4536. https://doi.org/10.3390/s24144536

Academic Editor: Angelo Maria Sabatini

Received: 2 May 2024
Revised: 12 June 2024
Accepted: 11 July 2024
Published: 13 July 2024

1. Introduction

One major challenge in sports science is identifying performance determinants to enhance coaching and competition outcomes [1]. Performance analysis plays a crucial role here, focusing on gathering valid, accurate, and reliable data during competitions to boost individual or team performance [2]. As one of the traditional methods of performance analysis in team sports, notational analysis seeks to obtain indicators of discrete actions and/or events by using advanced statistical procedures [3–5]. However, this method often fails to provide information regarding the dynamic confrontation of forces between the players and teams [6]. That is, teams continuously adjust and adapt their movement behaviour as a result of the cooperative (i.e., teammates' actions) and competitive interactions (i.e., opposition movements). This means that a team may dictate the game rhythm during the earlier phase of the match; however, as the match unfolds, it is likely that the opposing team will adjust their positioning and actions to balance the match [7]. In general, the discrete performance indicators captured by traditional notational analysis would fail to capture these coordinative tendencies between both teams [8]. As a result of recent technological developments, research in sports sciences started to analyse players' positioning dynamics, which considers the spatiotemporal relationships between both teams as a result of the collective principles of play, the opponents' behaviour, and the contextual circumstances [8–10]. Consequently, analysing players' positioning dynamics across the match seems to provide a more functional, holistic, and complex understanding of teams' sports performance.

In association football, performance analysis should be a comprehensive process involving precise measurements of physiological, technical, and tactical workloads that ultimately influence player and team outcomes [6]. The physical and physiological demands of the players when involved in real practice scenarios have been investigated incessantly over the last years, describing the movement patterns during training [11–13] and competition environments [14–16]. Nevertheless, these demands seem to be very sensitive to the teams' strategies, contextual variables, and opponent behaviour, indicating that multiple factors could impact players' physical responses during matches [1,17,18]. For example, lower external load has been reported in teams that show higher positioning synchronization during training sessions [19], shedding light on the role of positioning and tactical behaviour on the players' physical load. Positioning synchronization consists of a metric that measures the percentage of time that each pair of players moves in the same direction (e.g., the defensive line moving forward to follow the midfielders' and strikers' pressure) [9,19]. This variable has been used to distinguish teams' quality, as the winning team seems to possess higher values of movement synchronization [20]. More recently, rather than players' positioning, synchronization has been applied to players' movement speed. In this context, Gonçalves et al. [10] showed that higher dyadic synchronization at high speeds in the first half periods may limit players' performance in the second half. Accordingly, it was found a decrease in speed synchronization during the second half periods that may result from accumulated muscular and mental fatigue towards the match. Additionally, an examination of teams' behaviour across 15 min intervals during a single match revealed variations in team dispersion throughout these periods, with more regular patterns emerging toward the match's conclusion [7]. Altogether, the results from the previous study suggest that the integration of players' physical performance with the collective principles of play may be achieved by analysing the synchronization speed. Additionally, exploring the players' and team's performance across time periods for each half (e.g., blocks of 15 min) would contribute to a better understanding of their performance.

In fact, the analysis of positional dynamics aims to identify and describe emergent tactical patterns that underpin performance, while preserving the sequential and situational characteristics of match events [8,9,21,22]. This means that while analysing players' movements, researchers and data analysts must be aware that tactical patterns are dynamic and shift throughout the match, influenced by players' varying capacities and external factors such as pre-match coaching strategies that guide collective behaviour [23]. However, as the match unfolds, players and teams are likely to adapt to changing play configurations and opposition strategies, which are known as tactics [24]. Thus, analysing players' tactical performance during shorter periods can provide additional insights into how tactical decisions are executed under varying levels of fatigue [25,26].

Most research developed with positioning data has been applied to elite and adult levels. While studies examining youth players' tactical behaviour exist, they are predominantly centred around training sessions [27]. For example, Olthof, Frencken [28] compared under-13 (U13), under-15 (U15), under-17 (U17), and under-19 (U19) performance during small-sided games (SSGs) while varying the pitch size. The authors found that an increase in the pitch size contributed to a higher external load, and also bigger distances between teams [28]. In addition, higher variability was found in players' distances in larger formats [28]. This finding is especially important, as there has been a focus of discussion resulting from which size and playing format may be more appropriate for youth football players [29]. In fact, it is still common to find younger age groups (e.g., U13 and U15) playing 11-a-side in regular formats (e.g., length x width, 106 × 65 m playing area), which may not be appropriate for their development stage [29]. Despite the competitive setting concerns in youth football, research exploring their positional performance during matches is scarce. In fact, the limited available research exploring competitive formats in youth football has mostly compared it with SSGs [30]. Thus, exploring youth players' positioning performance across different time periods in competitive settings while comparing different age groups may help responsible bodies and entities to better frame competition for youth players. In addition, larger playing spaces seem to induce large variability in their behaviour [28]. It may also be expected to see a higher variability when playing during long periods (e.g., one half), while also resulting in lower tactical knowledge when compared to older levels [31]. Thus, this study aimed to explore how positional performance varies across different youth age groups (i.e., U13 and U15) and time periods during competitive matches.

2. Materials and Methods

2.1. Participants

The study encompassed 160 male outfield youth football players, with 80 participants U13 belonging to eight teams (U13: average age 12.5 ± 0.5 years; average height 163.2 ± 8.2 cm; average weight 48.9 ± 6.7 kg; average playing experience 4.3 ± 1.7 years) and 80 from U15, also belonging to eight teams (U15: average age 14.5 ± 0.5 years; average height 169.1 ± 9.5 cm; average weight 53.7 ± 7.1 kg; average playing experience 6.5 ± 1.4 years). The U13 teams engaged in three weekly training sessions (approximately 90 min each) and played an official 11-a-side game on weekends. Similarly, the U15 teams participated in four weekly training sessions (around 90 min each) and competed in an official 11-a-side game on weekends. Goalkeepers were involved in the study but excluded from data analysis due to their specialized positional constraints and unique game dynamics compared to outfield players. Informed consent was obtained from coaches, players, parents, and the club prior to the study's commencement. All participants were informed of their right to withdraw from the study at any time. The study's procedures were approved by the local Institutional Research Ethics Committee and conformed to the Declaration of Helsinki guidelines.

2.2. Procedures and Instruments

The teams involved in the study participated in eight official matches as part of the Second China Youth Football League 2023, with each age group (U13 and U15) playing four matches. The analysis focused on the 20 outfield players from each match's starting line-up who played the entire first half of each match. This approach was chosen because previous research has shown that player substitutions can significantly affect the tactical, physical, and technical performance of teams [32,33]. Given the high number of substitutions made by coaches during the second half, the study limited data analysis to the first half of each match to maintain consistency in the data collected and to minimize the impact of these changes on the analysis. Therefore, it was considered 35 min for U13 (an official match lasts for 70 min) and 40 min for U15 (an official match lasts for 80 min). The match sessions consisted of an 11 vs. 11 official match, on a 104 × 64 m pitch, with official rules. All players performed a 15 min standard warm-up consisting of ball possessing and dynamic stretching.

Before the beginning of each match, players were outfitted with a 10 Hz Catapult MinimaxX unit (MinimaxX S4, 10 Hz, Firmware 6.70, Catapult Innovations, Melbourne, Australia) which has been demonstrated to be valid and reliable [34]. The systems collected latitude and longitude coordinates, which were then extracted and resampled using an interpolation method to standardize the length of the time series. Subsequently, these coordinates were converted into metres using the Universal Transverse Mercator (UTM) coordinate system through specific coding routines [35]. The data were then smoothed with a 3 Hz Butterworth low-pass filter. To align the positional data with the field, a rotation matrix was applied, orienting the length of the playing field along the x-axis and the width along the y-axis. This matrix adjustment ensures that the players' positional data are consistent with the spatial orientation of the playing field, as detailed in the methodology outlined by Pereira, Gonçalves [36].

2.3. Positioning Relations

The positional data of the players were used to determine the following variables (see Figure 1)

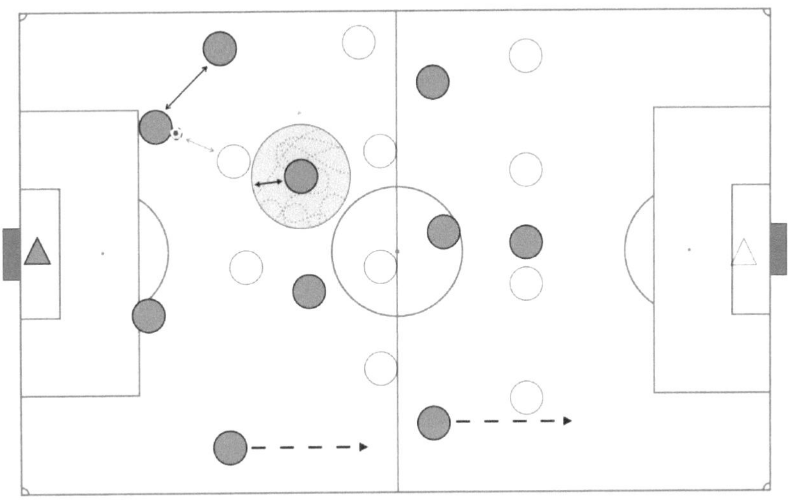

Figure 1. Representation of positional-related variables. Note: Dark grey circles represents one team, while light grey circles represents the other team.

:

- Frequency of near-in-phase synchronization from the players' speed displacements (expressed in % of time). Taking into consideration the all-possible intra-team dyads formed by the outfield teammates (45 dyads), the frequency of near-in-phase synchronization from the players' speed displacements was processed (expressed in % of time) [10]. The Hilbert Transform [37] was used to compute the relative phase of the time series corresponding to the speed displacements of all dyads. Near-in-phase synchronization (i.e., % of time spent between -30° and 30° of relative phase) was used to access players' interpersonal speed coordination.
- Spatial exploration index (SEI), which is processed by the calculation of the player's mean position and then computing all distances from this average point to all datasets

across the time series, and ending by computing the average value from all these distances [38].

- Distance to the nearest teammate and opponent expressed as absolute values (m), variability in these distances as expressed by the coefficient of variation (CV), and regularity in these distances expressed by the approximate entropy (ApEn) [39].

The ApEn has been used to assess the regularity in the players' movement behaviour, and its values range from 0 to 2 (arbitrary units). From a processing approach, ApEn expresses the probability that the configuration of one segment of the data in a time series will allow the prediction of the configuration of another segment of the time series a certain distance apart. In practice, this technique may be used, for example, to identify if players' positioning dynamics express a regular and predictable pattern which may, in turn, provide information regarding their tactical behaviour. The input values used to process the ApEn were 2 for the vector length (m) and $0.2 \times$ SD for the tolerance (r) [40,41].

2.4. Statistical Analysis

To evaluate variations in positional performance during matches, each match half analysed in the study was divided into three equal segments, or thirds, and this division was utilized as a factor in the analysis. Descriptive data were presented as means $\pm$ standard deviation (SD). Before inferential statistics, the Shapiro–Wilk and Levene's tests were performed to analyse whether the variables followed a normal distribution and verify the homogeneity of the variances, respectively. A two-way analysis of variance with repeated measures ANOVA [age group (U13 and U15) $\times$ half period (full, 1st, 2nd, and 3rd third)] was applied to test age and half period on the dependent variables. When significant main effects or interactions were achieved, Bonferroni post hoc analyses were performed to locate the pairwise. To estimate the strength of significant findings, effect sizes (ESs) were determined using Cohen's $d_{unbiased}$ [42,43]. Effect size values were interpreted as follows: <0.20 represents a trivial effect, 0.20 to 0.49 is classified as a small effect, 0.50 to 0.79 corresponds to an intermediate effect, and 0.80 and higher is considered a large effect [44]. The analysis reports the effect size using eta squared (η^2) for the main effects and interactions from the repeated measures ANOVA. For significant main effects or interactions, Cohen's $d_{unbiased}$ was used to indicate the effect size for the pairwise post hoc comparisons. The statistical analyses were conducted using SPSS software v.26 for Windows (IBM Corp., Armonk, NY, USA), and the significance level was established at $p \leq 0.05$.

3. Results

Table 1 presents the descriptive and inferential analysis for considered variables in both age groups and half periods. Figures 2–5 depict the descriptive result for visual inspection analysis, and Figures 6–9 depict the Cohen's $d_{unbiased}$ result for respective pairwise comparison.

Table 1. Descriptive and inferential analysis when comparing period effect (1st third × 2nd third; 1st third × 3rd third; and 2nd third × 3rd third), the age groups (U13 × U15), and also their interaction.

Variables	Age Group	Half Period				Repeated Measures Analysis								
						Period Effect (1st Third, 2nd Third, and 3rd Third)			Age Group Effect (U13 × U15)			Period * Age (Thirds × U13 and U15)		
		Full	1st Third	2nd Third	3rd Third	F	p	η^2_p	F	p	η^2_p	F	p	η^2_p
Speed synchronization % of time near-in-phase	U13	42.0 ± 6.5 $^{\#}$	45.9 ± 7.4 $^{a,b,\#}$	42.3 ± 7.7 $^{c,\#}$	37.8 ± 7.2 $^{\#}$	247.1	<0.001	0.26	142.3	<0.001	0.17	109.5	<0.001	0.13
	U15	36.1 ± 6.9	36.7 ± 7.4 ^{b}	36.5 ± 8.0 ^{c}	35.1 ± 7.6									
Spatial exploration index	U13	9.9 ± 2.2	11.1 ± 2.3 ^{a,b}	9.9 ± 2.2 $^{c,\#}$	7.7 ± 2.0 $^{\#}$	169.1	<0.001	0.52	2.2	0.14	0.01	27.3	<0.001	0.15
	U15	10.3 ± 1.6	11.4 ± 2.3 ^{a,b}	9.2 ± 1.7	9.3 ± 1.4									
Distance to the near teammate (nTM)														
Metres	U13	5.5 ± 1.1	5.5 ± 1.3	5.5 ± 1.4	5.6 ± 1.1	0.6	0.53	0.00	0.0	0.98	0.00	2.1	0.13	0.01
	U15	5.5 ± 1.2	5.7 ± 1.3 ^{b}	5.5 ± 1.3	5.4 ± 1.2									
Coefficient of variation %	U13	46.5 ± 7.4	48.0 ± 8.3 ^{a,b}	43.5 ± 7.8 $^{\#}$	44.1 ± 7.7	17.3	<0.001	0.10	1.4	0.24	0.01	7.4	<0.001	0.05
	U15	47.5 ± 6.4	47.7 ± 6.8 ^{b}	47.7 ± 8.9 ^{c}	43.4 ± 7.6									
Approximate entropy	U13	0.078 ± 0.020	0.082 ± 0.020 ^{b}	0.083 ± 0.023 ^{c}	0.075 ± 0.021	26.7	<0.001	0.14	0.2	0.65	0.00	4.8	0.01	0.03
	U15	0.076 ± 0.017	0.085 ± 0.021 ^{a,b}	0.076 ± 0.017 ^{c}	0.072 ± 0.019									
Distance to the near Opponent (nOPP)														
Metres	U13	4.2 ± 1.1	5.0 ± 1.3 $^{a,b,\#}$	4.2 ± 1.2 ^{c}	3.5 ± 1.2 $^{\#}$	4.9	0.01	0.03	0.9	0.35	0.01	293.3	<0.001	0.65
	U15	4.4 ± 1.3	3.6 ± 1.2 ^{a,b}	4.3 ± 1.5 ^{c}	5.4 ± 1.5									
Coefficient of variation %	U13	60.8 ± 10.3	56.1 ± 9.9 ^{b}	58.6 ± 11.6	60.9 ± 13.8	19.7	<0.001	0.11	1.5	0.22	0.01	1.9	0.16	0.01
	U15	63.0 ± 8.3	55.5 ± 7.6 ^{a,b}	61.6 ± 11.4	63.3 ± 11.8									
Approximate entropy	U13	0.081 ± 0.027	0.090 ± 0.031 ^{b}	0.084 ± 0.033 $^{\#}$	0.081 ± 0.032	24.9	<0.001	0.14	1.4	0.25	0.01	5.1	0.01	0.03
	U15	0.075 ± 0.026	0.094 ± 0.034 ^{a,b}	0.074 ± 0.028	0.074 ± 0.031									

Post hoc analysis—period * age. $^{\#}$ U13 full vs. U15 full; U13 1st third vs. U15 1st third; U13 2nd third vs. U15 2nd third; U13 3rd third vs. U15 3rd third. ^{a} 1st third vs. 2nd third. ^{b} 1st third vs. 3rd third. ^{c} 2nd third vs. 3rd third.

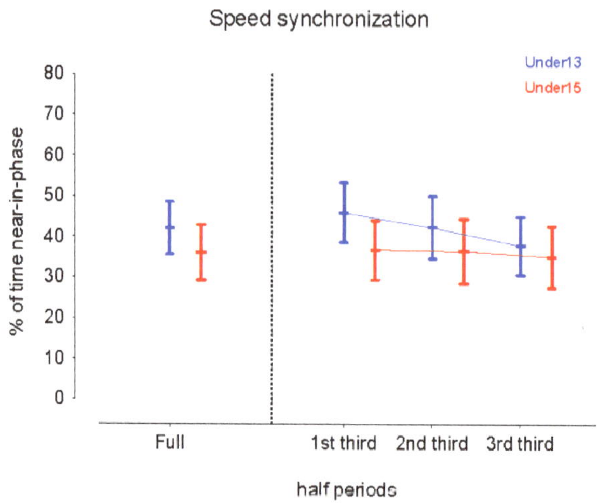

Figure 2. Descriptive values for players' speed displacement synchronization according to the game periods (1st third, 2nd third, and 3rd third) and age groups (U13 and U15). Each dot represents an intra-team dyad value and the coloured error bars indicate mean ± standard deviation.

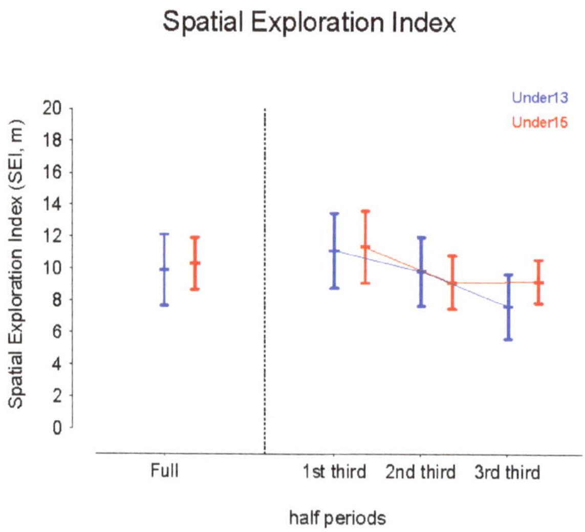

Figure 3. Descriptive values for the players' spatial exploration index (SEI) according to the game periods (1st third, 2nd third, and 3rd third), age groups (U13 and U15), and their interactions. Each dot represents an intra-team dyad value and the coloured error bars indicate mean ± standard deviation.

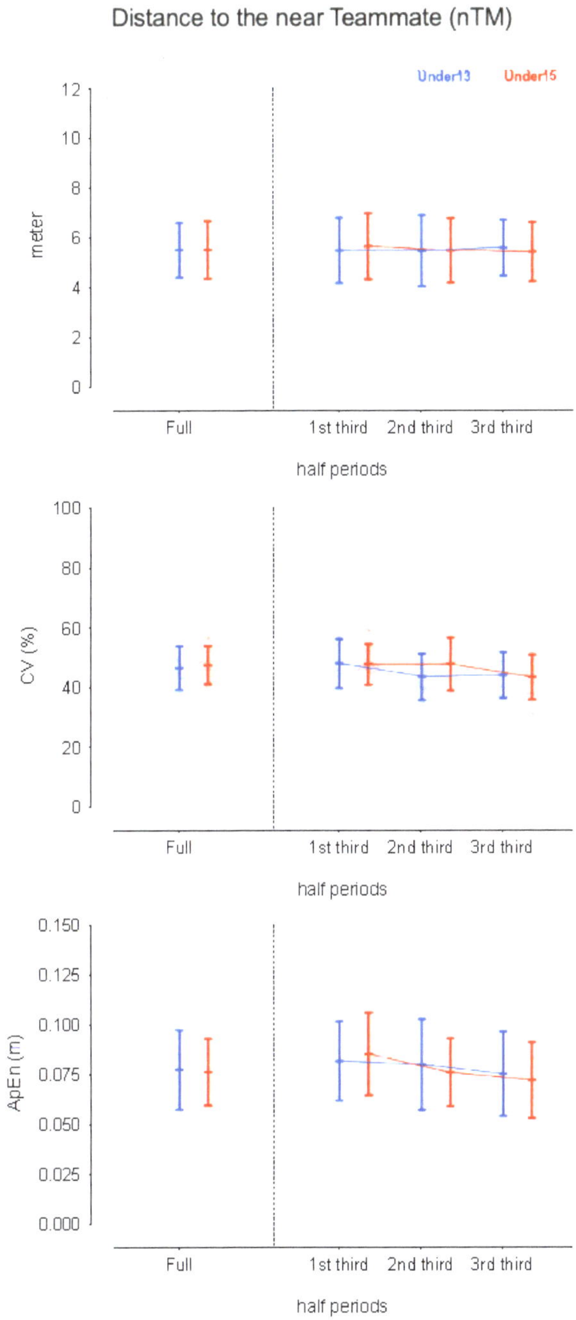

Figure 4. Descriptive values for players' distance to the near teammate (nTM) according to the game periods (1st third, 2nd third, and 3rd third), age groups (U13 and U15), and their interactions. Each dot represents an intra-team dyad value and the coloured error bars indicate mean ± standard deviation. CV = coefficient of variation; ApEn = approximate entropy.

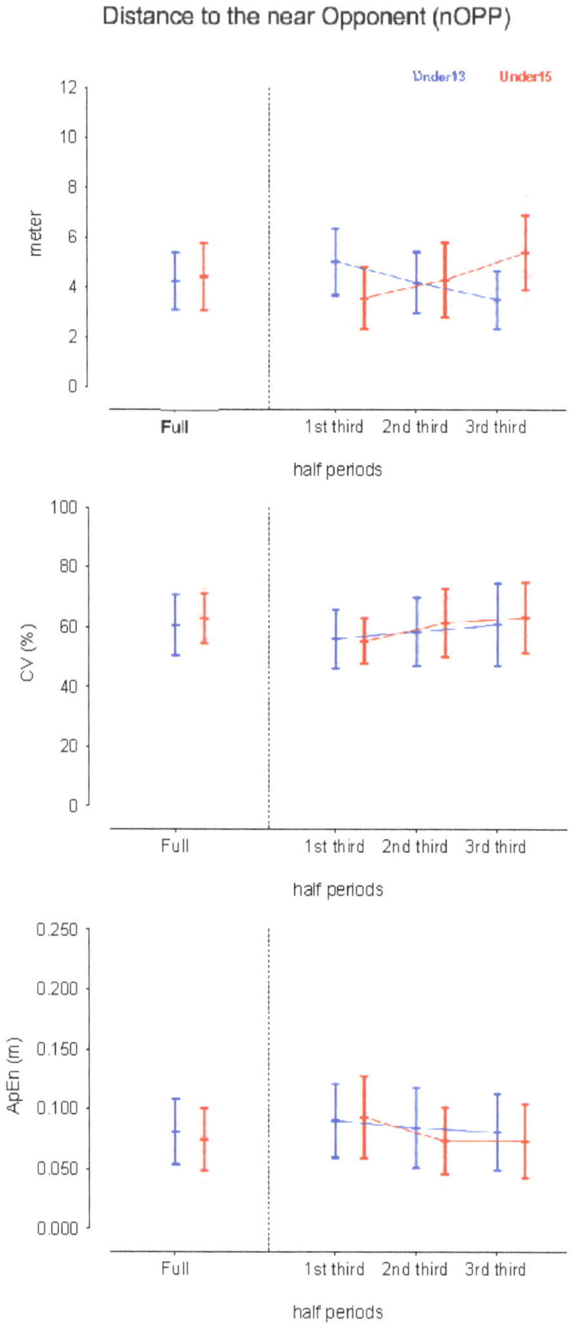

Figure 5. Descriptive values for players' distance to the near opponent (nOPP) according to the game periods (1st third, 2nd third, and 3rd third), age groups (U13 and U15), and their interactions. Each dot represents an intra-team dyad value and the coloured error bars indicate mean ± standard deviation. CV = coefficient of variation; ApEn = approximate entropy.

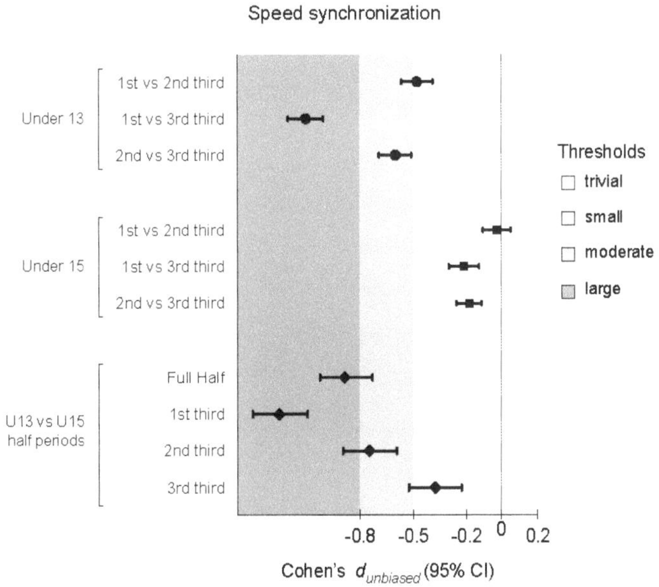

Figure 6. Cohen's $d_{unbiased}$ differences for players' speed displacement synchronization according to the game periods (1st third, 2nd third, and 3rd third), age groups (U13 and U15), and their interactions. Error bars indicate uncertainty in the true mean changes with 95% confidence intervals.

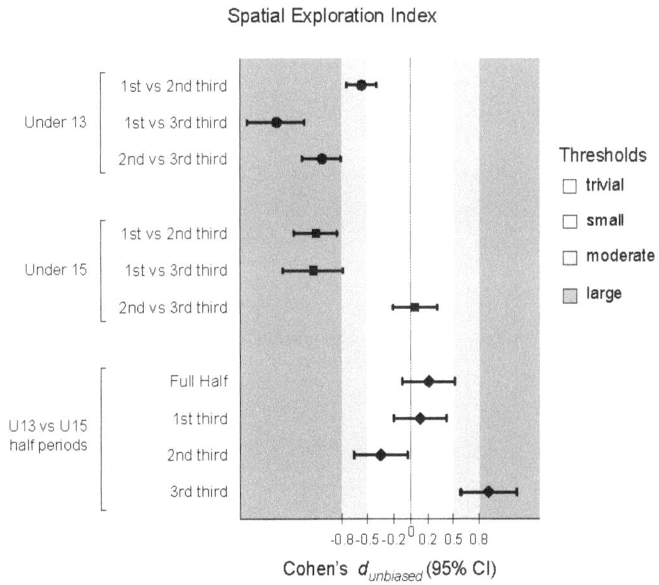

Figure 7. Cohen's *d* differences for the players' spatial exploration index according to the game periods (1st third, 2nd third, and 3rd third), age groups (U13 and U15), and their interactions. Error bars indicate uncertainty in the true mean changes with 95% confidence intervals.

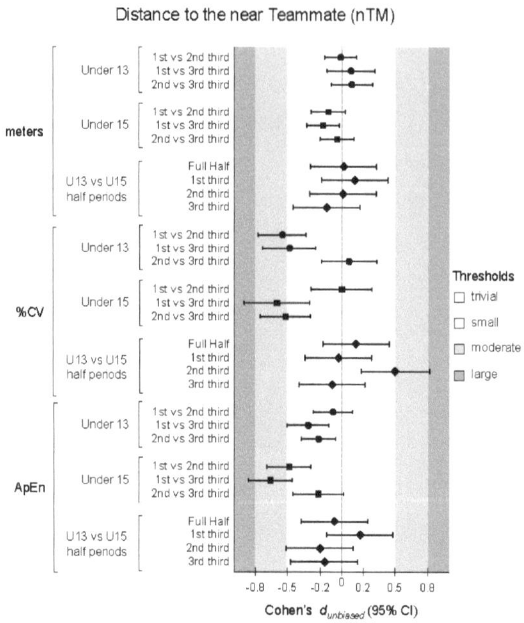

Figure 8. Cohen's *d* differences for players' distance to the near teammate (nTM) according to the game periods (1st third, 2nd third, and 3rd third), age groups (U13 and U15), and their interactions. Error bars indicate uncertainty in the true mean changes with 95% confidence intervals. CV = coefficient of variation; ApEn = approximate entropy.

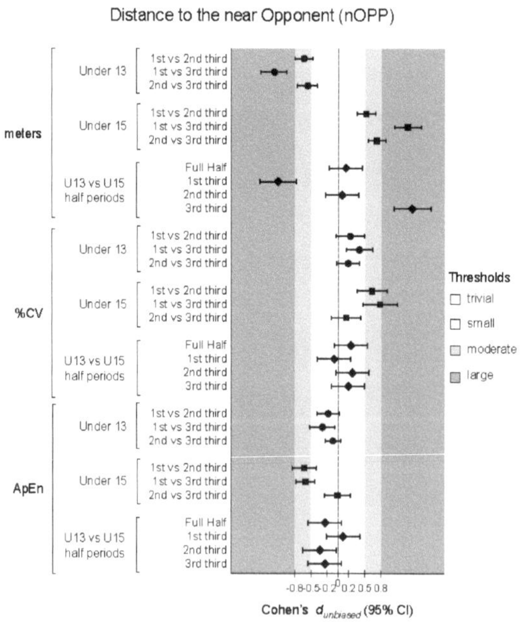

Figure 9. Cohen's *d* differences for players' distance to the near opponent (nOPP) according to the game periods (1st third, 2nd third, and 3rd third), age groups (U13 and U15), and their interactions. Error bars indicate uncertainty in the true mean changes with 95% confidence intervals. CV = coefficient of variation; ApEn = approximate entropy.

The analysis commenced by examining the interaction between the half period (with three levels: 1st third, 2nd third, and 3rd third) and age group (U13 vs. U15) on % of the time in near-in-phase speed synchronization. A two-way ANOVA with repeated measures revealed a significant interaction effect, with F = 109.5, $p < 0.001$, and $\eta^2_p = 0.13$, indicating that the synchronization over time differed between the groups. The main effect of the half period was also significant, with F = 247.1, $p < 0.001$, and $\eta^2_p = 0.26$, suggesting that the % of synchronization changed over time, regardless of the group assignment. Additionally, the main effect of the group was significant, with F = 142.3, $p < 0.001$, and $\eta^2_p = 0.17$. This suggests that, overall, the U15 group exhibited a lower % of synchronization across all time points compared to the U13 group (see Table 1). Considering the post hoc analysis as well as Cohen's $d_{unbiased}$ results, and comparing U13 vs. U15, U13 had significantly ($p < 0.001$) more % of synchronization in all considered half periods: a large effect for both the full half (Cohen $d_{unbiased}$ [95% CI]; −0.89 [−1.03; −0.73]) and the 1st third (−1.26 [−1.41; −1.10]); a moderate effect for the 2nd third (−0.75 [−0.91; −0.61]); and a small effect for the 3rd third (−0.38 [−0.53; −0.23]). The U13 period significantly decreased over the half while trivial to small results were identified for the U15 period in comparison (see Figures 2 and 3).

The players' SEI showed a significant effect on the period*age interaction, F = 27.3, $p < 0.001$ and $\eta^2_p = 0.15$, and the half period, with F = 169.1, $p < 0.001$, and $\eta^2_p = 0.52$ (see Table 1). U13 decreased their values over the match, from a small to large effect size, while U15 decreased only after the 1st third (1st vs. 2nd third: −1.10 [1.37; −0.87], and 1st vs. 3rd third: −1.13 [−1.49; −0.79], both large effect sizes). On the 3rd third, U15 showed higher values compared to U13 (see Figures 4 and 5).

The players' distance to the near teammate nTM was analysed from absolute values, metres, the coefficient of variation (%CV) as the magnitude of the variability, and approximate entropy (ApEn) as the magnitude of the structure variability. The absolute values were similar for both age groups and for all periods (see Table 1). However, the %CV showed significant differences in the period*age interaction, F = 7.4, $p < 0.001$ and $\eta^2_p = 0.05$, and the half period, with F = 17.3, $p < 0.001$, and $\eta^2_p = 0.10$ (see Table 1). The U13 group decreased from the 1st to 2nd third (1st vs. 2nd third: −0.55 [−0.77; −0.33]; and 1st vs. 3rd third: −0.48 [−0.73; −0.24]), while U15 decreased from the 2nd to 3rd third (1st vs. 3rd third: −0.60 [−0.90; −0.30]; and 2nd vs. 3rd third: −0.52 [−0.76; −0.29]). The ApEn also presented significant differences in the period*age interaction, F = 4.8, $p = 0.01$ and $\eta^2_p = 0.03$, and the half period, F = 26.7, $p < 0.001$ and $\eta^2_p = 0.14$ (see Table 1). Pairwise differences showed that U13 and U15 presented similar values. However, for U13, the 1st and 2nd third were similar, and the distance to nTM became more regular in the 3rd third (1st vs. 3rd third: −0.31 [−0.50; −0.12]; and 2nd vs. 3rd third: −0.21 [−0.37; −0.05]). U15 decreased the ApEn value across the match (see Figures 6 and 7).

The players' distance to the near opponent (nOPP), considering absolute values, revealed a significant interaction effect, with F = 293.3, $p < 0.001$, and $\eta^2_p = 0.65$, indicating that the nOPP over time differed between the groups and match period, with F = 4.9, $p = 0.01$, and $\eta^2_p = 0.06$, suggesting that nOPP changed over time (see Table 1). Additionally, while the main effect of the group was not significant, the pairwise differences presented a lower distance to nOPP for U15 in the 1st third (−1.11 [−1.45; −0.78]) and higher values in the 3rd third (1.39 [1.05; 1.74]). In fact, U13 decreased the distance to nOPP over the match while U15 increased during the same period (moderate to large effect size for both age groups). The %CV only revealed a significant half-period effect, with F = 19.7, $p < 0.001$, and $\eta^2_p = 0.11$, where the values of nOPP increased over time for both groups (see Table 1). Finally, the ApEn presented significant differences in the period * age interaction, with F = 5.1, $p = 0.01$, and $\eta^2_p = 0.03$, and the half period, with F = 29.9, $p < 0.001$, and $\eta^2_p = 0.14$ (see Table 1). Pairwise differences showed that U13 and U15 presented similar values. However, for U15, the distance to nOPP becomes more regular right after the 1st third (1st vs. 2nd third: −0.62 [−0.85; −0.40]; and 1st vs. 3rd third: −0.60 [−0.77; −0.43]) (see Figures 7 and 8).

4. Discussion

This study aimed to explore and compare changes in positioning performance among youth soccer players (U13 and U15) during an 11-a-side match. Generally, results from the U13 group indicated a decrease in synchronization speed and the amount of space explored, along with a reduction in distance to the nearest opponent (nOPP). For the U15 group, differences were primarily observed in the third period of the match, with a decrease in synchronization speed and in SEI, while the distance and regularity to the nOPP increased. When comparing both age groups, differences in synchronization speed were noted throughout the entire match, as well as in all periods, with higher values being observed in the U13 group. Additionally, differences between age groups became more pronounced as the game progressed, particularly in the second and third periods.

4.1. Analysis of U13 Positioning Variation across Time Periods

A major aim with younger age groups (i.e., from U5 to U14) is to develop players' technical and coordination skills [45–47] while developing players' understanding of the general (i.e., reject numerical inferiority, avoid numerical equality, and seek numerical superiority) and specific principles of play (i.e., offensive and defensive behaviours that guide individual, group, and collective movement behaviours) [48]. Developing such skills seems to be a determinant for future achievements in football competitive environments [49,50]. The development of such technical, coordinative, and tactical skills must be grounded in learning environments that foster decision-making skills, and competitive and cooperative interactions. In fact, decision-making and proper positioning seem to be related to talent in football [51]. Therefore, a meticulous and careful long-term plan is required to enhance the chances of youth players to progress in football. In line with this, a high number of football associations and researchers have been exploring which competitive formats may be more suitable for the different age groups [29,52]. For example, Sanchez, Ramirez-Campillo [53] compared U12 performance in 7-a-side, 8-a-side, and 11-a-side conditions and found higher external load in the larger format compared to the other two conditions. From the technical perspective, the seven-a-side format seems to elicit a greater number of actions when compared to the eight-a-side format in U12 [54]. A similar trend was found by Joo, Hwang-Bo [30] who explored the effects of using SSGs (8-a-side) in smaller (length × width, 68 × 47 m) or regular spaces (75 × 47 m) when compared to official matches (11-a-side, 75 × 47 m) in U12 Korean players. Altogether, the results of these studies seem to highlight that the 11-a-side format may be significantly complex for U12 players, who may not possess the technical (e.g., long pass ability) nor decision-making skills (i.e., the ability to scan the environment to perceive teammates' and opponents' positioning) that may allow them to successfully perform in such designs.

A similar conclusion may be drawn from the present study in the U13 age groups when analysing their tactical behaviour. That is, there was a decrease in synchronization speed, SEI, and distance to nOPP, while there was an increase in the regularity of the distance to the nearest opponent across the half thirds. These results suggest that as the match unfolds, there is a shift in the players' focus from the collective movement behaviour towards the direct opponent. In fact, from the 1st third towards the 3rd third, there is a decrease of almost 1.5 m in the distance to the nOPP, which was followed by an increase in the regularity of this distance. In other words, players seem to become closer to their direct opponent, while maintaining this distance across the half-periods. Accordingly, younger age groups seem to be more focused on the ball and on the closest opponent than on the team's collective approach [55], which may justify these results. Interestingly, these trends were more evident across thirds, suggesting that players were able to keep a collective strategy for the match during the first 15 min. Although anecdotally, as the coach's instruction was not measured, the pre-match speeches are often focused on providing descriptions of players' roles, emphasizing information about the opposition's weakness, while providing information on how to collectively behave during the different game phases [23]. Thus, it may be plausible to assume that U13 players are able to follow a collective strategy within the first minutes of the match, after which it seems to fade into a more individual focus on the ball and the opponent. In fact, younger age groups attempt to solve the game problems by adopting an individual approach rather than a collective one [27].

From a practical point of view, governmental entities and national football associations must consider the type of competitive designs in youth age groups. For example, smaller formats may be more suitable for the U13 group. Alternatively, it may be important to add stoppage periods that may allow coaches to provide individual and collective feedback, allowing the players to adjust their tactical behaviours.

4.2. *Analysis of U15 Positioning Variation across Time Periods*

Older age group players seem to be more able to move and adjust to the competitive environment [56] by being able to identify the relevant information to unfold goal-directed behaviours as a result of better perceptual and cognitive skills [57]. In general, 11 vs. 11 formats are used from the U14 age groups above across different countries [29], which may suggest that this age is a point at which players might be able to perceive and act within complex competitive environments. The results from the present study seem to support this statement, as the U15 positional variables (e.g., SEI, distance to the nearest teammate, and ApEn in the distance to the nearest opponent) seem to be less affected across half-period thirds. For instance, most variations in players' performance emerge in the 3rd third, with decreases in speed synchronization, SEI, and distance to the nTM, lower variability, and higher regularity in the distance to the NTM. In contrast, a bigger distance toward the nOPP was found. In general, these results point out that U15 can keep its performance constant for most variables across the first two thirds of the half. Based on this information, the transition to the 11 vs. 11 format may require a rest period around the middle of each half that may allow the players to reorganize their positioning. Still, a different strategy is depicted when compared to the U13 age group. That is, while in the U13 group, a decrease in the distance to the nOPP was found, an opposite trend was identified for the U15 group. Thus, it seems that with increased fatigue resulting from the competitive interactions, U15 adopts a more collective approach by decreasing the distance to the nTM and increasing it towards the nOPP. These findings are in line with the study of Coutinho, Gonçalves [58], who explored how U14 players' positioning performance was affected during small-sided games by performing with additional muscular fatigue. The authors found a decrease in the distance between dyads, while also observing greater movement coordination. In addition, in this study, it was also found that there was a lower variation and higher regularity in the distance to the nTM for the U15 age group from the 1st to the 3rd third. A previous study showed higher values for the inter-team distance in the U15 age group than in the U13 group, which may reinforce these results. In contrast, a higher coefficient of variation in the nearest was found in both the 2nd and 3rd third when compared to the 1st third. This variability may act as a functional movement behaviour, because of the higher compactness (i.e., expressed by the lower distance to the nTM and SEI). In fact, variability in players' movement behaviours has been considered fundamental to adjusting to the dynamic and unpredictable nature of competitive football settings [59].

4.3. *Differences between U13 and U15*

A wide body of research has been exploring differences between age groups from a tactical point of view. For example, Folgado, Lemmink [27] compared the performance of U9, U11, and U13 under three-a-side and four-a-side small-sided game formats. The three-a-side format revealed major differences in the distance between players, where the older players revealed a greater ability to use the pitch length, while similar distances between players were identified for the four-a-side format. Olthof, Frencken [28] compared U13, U15, U17, and U19 performances during five-a-side small-sided games while varying pitch dimensions (i.e., small, 40 × 30 m; and large, 68 × 47 m). The results showed a greater distance between players and the playing area in the U15 group when compared to the U13 group. The same trend was identified by a recent study comparing U13, U15, and U18 players' positioning performance during a five-a-side small-sided game [31]. Older players revealed larger areas, and also a bigger distance between teams. The combined findings from these studies highlight that older players are more able to use the available space. In the present study, major differences between age groups were identified for speed synchronization and SEI. In this respect, higher values of synchronization speed were identified in all thirds for the U13 group. As previously noted, there is a higher trend towards following the ball movement in younger age groups [55], which may have contributed to such values. That is, this age group seems to be less prone to move collectively, but rather, they focus on the ball movement, and thus, it may be expected that both teams move as a result of the ball's location. In contrast, the U15 group may possess higher tactical awareness that allows them to vary between moving collectively (e.g., staying compact while defending to press the opposition) or moving at different paces, rhythms, and directions (e.g., attempting to perform depth passes in the last third, whereas, one to two players may move close to the ball to drag defenders, with one or two sprinting to explore the space). Thus, the lower synchronization values in the U15 group may reflect this age group's ability to understand each configuration of play. In fact, this group was less affected by the thirds. For example, the U13 group showed a clear trend towards decreasing the space explored as the thirds progressed, while the U15 group despite decreasing from the 1st third to the 2nd, was kept constant to the 3rd third. Younger age groups, such as the U13 group, are likely to adopt more individual strategies to solve game problems than explore collective movement solutions [27]. Consequently, and as the match unfolds, they may decrease the space exploration as a result of the lower collective commitment. In contrast, the U15 group revealed a decrease from the 1st to the 2nd third but kept the values constant to the 3rd third. The values from the 1st third may result from the inherent variability in team behaviours in the first 15 min, in which both teams may be exploring adaptive movement patterns [7]. However, as the match unfolds and the fatigue increases, U15 players may adopt more collective and stable behaviours [58].

5. Conclusions

Overall, it is important to be aware that exposing young players to 11 vs. 11 matches for long periods may not provide an appropriate learning environment, especially in the U13 age group. The high density of players and available space contributed to more variable and irregular behaviours across time, which can be depicted from the lower speed displacement % synchronization (i.e., collective variable) and higher SEI (i.e., individual variable). In contrast, the U15 group appears to be able to reveal positional adjustments over time, reflecting their higher tactical awareness. These findings highlight the necessity of tailoring youth football training and competition structures to suit the developmental needs and capabilities of various age groups, thereby optimizing learning and performance outcomes.

Author Contributions: Conceptualization, Q.L. and Z.H.; methodology, Q.L., X.W. and T.A.; software, Q.L. and B.G.; formal analysis, Q.L., D.C. and B.G.; resources, Q.L. and Z.H; writing—original draft preparation, Q.L., D.C. and B.G.; writing—review and editing, all authors; visualization, Q.L. and B.G.; supervision, Z.H. and B.G.; project administration, Z.H.; funding acquisition, Z.H. All authors have read and agreed to the published version of the manuscript.

Funding: This research received no external funding.

Institutional Review Board Statement: The study was conducted in accordance with the Declaration of Helsinki and approved by the Institutional Ethics Committee of Beijing Sport University.

Informed Consent Statement: Informed consent was obtained from all subjects involved in the study.

Data Availability Statement: In order to protect the subjects' confidentiality and privacy, data are only available upon request. Interested researchers may contact the corresponding author.

Conflicts of Interest: The authors declare no conflicts of interest.

References

1. McGarry, T. Applied and theoretical perspectives of performance analysis in sport: Scientific issues and challenges. *Int. J. Perform. Anal. Sport* **2009**, *9*, 128–140. [CrossRef]
2. O'Donoghue, P. *An Introduction to Performance Analysis of Sport*; Routledge: London, UK, 2014; pp. 26–56.
3. Hughes, M.; Franks, I. *Notational Analysis of Sport*, 2nd ed.; Routledge: London, UK, 2004.
4. Carling, C.; Reilly, T.; Williams, A.M. *Performance Assessment for Field Sports*; Routledge: London, UK; New York, NY, USA, 2009.
5. Hughes, M.; Franks, I. *The Essentials of Performance Analysis: An Introduction*; Taylor & Francis: London, UK, 2007.
6. Travassos, B.; Davids, K.; Araújo, D.; Esteves, T.P. Performance analysis in team sports: Advances from an Ecological Dynamics approach. *Int. J. Perform. Anal. Sport* **2013**, *13*, 83–95. [CrossRef]
7. Duarte, R.; Araújo, D.; Folgado, H.; Esteves, P.; Marques, P.; Davids, K. Capturing complex, non-linear team behaviours during competitive football performance. *J. Syst. Sci. Complex.* **2013**, *26*, 62–72. [CrossRef]
8. Duarte, R.; Araújo, D.; Correia, V.; Davids, K.; Marques, P.; Richardson, M. Competing together: Assessing the dynamics of team–team and player–team synchrony in professional association football. *Hum. Mov. Sci.* **2013**, *32*, 555–566. [CrossRef] [PubMed]
9. Folgado, H.; Duarte, R.; Fernandes, O.; Sampaio, J. Competing with lower level opponents decreases intra-team movement synchronization and time-motion demands during pre-season soccer matches. *PLoS ONE* **2014**, *9*, e97145. [CrossRef] [PubMed]
10. Gonçalves, B.; Coutinho, D.; Travassos, B.; Folgado, H.; Caixinha, P.; Sampaio, J. Speed synchronization, physical workload and match-to-match performance variation of elite football players. *PLoS ONE* **2018**, *13*, e0200019. [CrossRef] [PubMed]
11. Dellal, A.; Owen, A.; Wong, D.P.; Krustrup, P.; Van Exsel, M.; Mallo, J. Technical and physical demands of small vs. large sided games in relation to playing position in elite soccer. *Hum. Mov. Sci.* **2012**, *31*, 957–969. [CrossRef] [PubMed]
12. Krustrup, P.; Mohr, M.; Ellingsgaard, H.; Bangsbo, J. Physical demands during an elite female soccer game: Importance of training status. *Med. Sci. Sport. Exerc.* **2005**, *37*, 1242–1248. [CrossRef]
13. Hill-Haas, S.; Dawson, B.; Impellizzeri, F.; Coutts, A. Physiology of small-sided games training in football a systematic review. *Sport. Med.* **2011**, *41*, 199–220. [CrossRef]
14. Carling, C.; Dupont, G. Are declines in physical performance associated with a reduction in skill-related performance during professional soccer match-play? *J. Sport. Sci.* **2011**, *29*, 63–71. [CrossRef]
15. Di Salvo, V.; Baron, R.; Tschan, H.; Calderon Montero, F.J.; Bachl, N.; Pigozzi, F. Performance characteristics according to playing position in elite soccer. *Int. J. Sport. Med.* **2007**, *28*, 222–227. [CrossRef] [PubMed]
16. Lago, C.; Casais, L.; Dominguez, E.; Sampaio, J. The effects of situational variables on distance covered at various speeds in elite soccer. *Eur. J. Sport Sci.* **2010**, *10*, 103–109. [CrossRef]
17. Travassos, B.; Araújo, D.; Davids, K.; O'Hara, K.; Leitão, J.; Cortinhas, A. Expertise effects on decision-making in sport are constrained by requisite response behaviours—A meta-analysis. *Psychol. Sport Exerc.* **2013**, *14*, 211–219. [CrossRef]
18. Sampaio, J.; Lago, C.; Gonçalves, B.; Maçãs, M.; Leite, N. Effects of pacing, status and unbalance in time motion variables, heart rate and tactical behaviour when playing 5-a-side football small-sided games. *J. Sci. Med. Sport* **2014**, *17*, 229–233. [CrossRef] [PubMed]
19. Folgado, H.; Gonçalves, B.; Sampaio, J. Positional synchronization affects physical and physiological responses to preseason in professional football (soccer). *Res. Sport. Med.* **2018**, *26*, 51–63. [CrossRef] [PubMed]
20. Folgado, H.; Duarte, R.; Marques, P.; Gonçalves, B.; Sampaio, J. Exploring how movement synchronization is related to match outcome in elite professional football. *Sci. Med. Footb.* **2018**, *2*, 101–107. [CrossRef]
21. Ric, A.; Torrents, C.; Gonçalves, B.; Sampaio, J.; Hristovski, R. Soft-assembled Multilevel Dynamics of Tactical Behaviors in Soccer. *Front. Psychol.* **2016**, *7*, 1513. [CrossRef] [PubMed]
22. Raya-Castellano, P.E.; García Calvo, T.; López-Gajardo, M.A.; Marzano-Felisatti, J.M.; Pulido, J.J. The pre-match talk: A multiple-method examination of youth-elite football coaches' motivational content. *Int. J. Sport. Sci. Coach.* **2024**, *19*, 604–619. [CrossRef]
23. Hibbs, A.; O'Donoghue, P. Strategy and Tactics in Sports Performance. In *Routledge Handbook of Sports Performance Analysis*; Routledge: London, UK, 2013; pp. 248–258.
24. Grehaigne, J.; Godbout, P. Tactical Knowledge in Team Sports From a Constructivist and Cognitivist Perspective. *QUEST* **1995**, *47*, 490–505. [CrossRef]
25. Mohr, M.; Krustrup, P.; Bangsbo, J. Match performance of high-standard soccer players with special reference to development of fatigue. *J. Sport. Sci.* **2003**, *21*, 519–528. [CrossRef]
26. Carling, C.; Le Gall, F.; Dupont, G. Are physical performance and injury risk in a professional soccer team in match-play affected over a prolonged period of fixture congestion? *Int. J. Sport. Med.* **2012**, *33*, 36–42. [CrossRef] [PubMed]
27. Folgado, H.; Lemmink, K.A.; Frencken, W.; Sampaio, J. Length, width and centroid distance as measures of teams tactical performance in youth football. *Eur. J. Sport Sci.* **2014**, *14* (Suppl. S1), S487–S492. [CrossRef] [PubMed]
28. Olthof, S.B.H.; Frencken, W.G.P.; Lemmink, K.A.P.M. Match-derived relative pitch area changes the physical and team tactical performance of elite soccer players in small-sided soccer games. *J. Sport. Sci.* **2018**, *36*, 1557–1563. [CrossRef] [PubMed]
29. Brito, Â.M.P.; Maia, J.A.R.; Garganta, J.M.; Duarte, R.F.L.; Diniz, A.M.F.A. The game variants in Europe. Trends and perspectives during youth competitive stages. *Motriz Rev. Educ. Física* **2017**, *23*, e101753. [CrossRef]

30. Joo, C.H.; Hwang-Bo, K.; Jee, H. Technical and Physical Activities of Small-Sided Games in Young Korean Soccer Players. *J. Strength Cond. Res.* **2016**, *30*, 2164–2173. [CrossRef] [PubMed]
31. Clemente, F.M.; Castillo, D.; Los Arcos, A. Tactical Analysis According to Age-level Groups during a 4 vs. 4 Plus Goalkeepers Small-sided Game. *Int. J. Environ. Res. Public Health* **2020**, *17*, 1667. [CrossRef] [PubMed]
32. Lorenzo-Martínez, M.; Rein, R.; Garnica-Caparrós, M.; Memmert, D.; Rey, E. The Effect of Substitutions on Team Tactical Behavior in Professional Soccer. *Res. Q. Exerc. Sport.* **2022**, *93*, 301–309. [CrossRef] [PubMed]
33. Lorenzo-Martínez, M.; Padrón-Cabo, A.; Rey, E.; Memmert, D. Analysis of Physical and Technical Performance of Substitute Players in Professional Soccer. *Res. Q. Exerc. Sport.* **2021**, *92*, 599–606. [CrossRef] [PubMed]
34. Johnston, R.J.; Watsford, M.L.; Kelly, S.J.; Pine, M.J.; Spurrs, R.W. Validity and interunit reliability of 10 Hz and 15 Hz GPS units for assessing athlete movement demands. *J. Strength Cond. Res.* **2014**, *28*, 1649–1655. [CrossRef]
35. Palacios, R. deg2utm. MATLAB Central File Exchange. Available online: http://www.mathworks.com/matlabcentral/fileexchange/10915 (accessed on 20 October 2016).
36. Pereira, P.; Gonçalves, B.; Bravo, J.; Fernandes, O.; Folgado, H. Multidimensional Variables. In *The Use of Applied Technology in Team Sport*; Pino-Ortega, J., Rico-González, M., Eds.; Routledge: New York, NY, USA, 2021; p. 176.
37. Palut, Y.; Zanone, P.G. A dynamical analysis of tennis: Concepts and data. *J. Sport. Sci.* **2005**, *23*, 1021–1032. [CrossRef]
38. Gonçalves, B.; Esteves, P.; Folgado, H.; Ric, A.; Torrents, C.; Sampaio, J. Effects of Pitch Area-Restrictions on Tactical Behavior, Physical, and Physiological Performances in Soccer Large-Sided Games. *J. Strength Cond. Res.* **2017**, *31*, 2398–2408. [CrossRef] [PubMed]
39. Gonçalves, B.; Marcelino, R.; Torres-Ronda, L.; Torrents, C.; Sampaio, J. Effects of emphasising opposition and cooperation on collective movement behaviour during football small-sided games. *J. Sport. Sci.* **2016**, *34*, 1346–1354. [CrossRef] [PubMed]
40. Pincus, S.M. Approximate entropy as a measure of system complexity. *Proc. Natl. Acad. Sci. USA* **1991**, *88*, 2297–2301. [CrossRef] [PubMed]
41. Yentes, J.M.; Hunt, N.; Schmid, K.K.; Kaipust, J.P.; McGrath, D.; Stergiou, N. The appropriate use of approximate entropy and sample entropy with short data sets. *Ann. Biomed. Eng.* **2013**, *41*, 349–365. [CrossRef] [PubMed]
42. Ho, J.; Tumkaya, T.; Aryal, S.; Choi, H.; Claridge-Chang, A. Moving beyond P values: Data analysis with estimation graphics. *Nat. Methods* **2019**, *16*, 565–566. [CrossRef] [PubMed]
43. Cumming, G.; Calin-Jageman, R. *Introduction to the New Statistics: Estimation, Open Science, and Beyond*; Routledge: New York, NY, USA, 2024.
44. Cohen, J. *Statistical Power Analysis for the Behavioral Sciences*, 2nd ed.; L. Erlbaum Associates: Hillsdale, NJ, USA, 1988.
45. Gonzalez-Villora, S.; García-López, L.M.; Gutiérrez-Díaz, D.; Pastor-Vicedo, J.C. Tactical awareness, decision making and skill in youth soccer players (under-14 years). *J. Hum. Sport Exerc.* **2013**, *8*, 412–426. [CrossRef]
46. Kokstejn, J.; Musálek, M. The Relationship between Fundamental Motor Skills and Game Specific Skills in Elite Young Soccer Players. *J. Phys. Educ. Sport* **2019**, *19*, 249.
47. Stafford, I. *Coaching for Long-term Athlete Development: To Improve Participation and Performance in Sport*; Coachwise Business Solutions on behalf Sports Coach UK: Leeds, UK, 2005.
48. Ouellette, J. Principles of Play for Soccer. *Strateg. A J. Phys. Sport Educ.* **2004**, *17*, 26.
49. Forsman, H.; Blomqvist, M.; Davids, K.; Liukkonen, J.; Konttinen, N. Identifying technical, physiological, tactical and psychological characteristics that contribute to career progression in soccer. *Int. J. Sport. Sci. Coach.* **2016**, *11*, 505–513. [CrossRef]
50. Huijgen, B.C.H.; Elferink-Gemser, M.T.; Post, W.J.; Visscher, C. Soccer Skill Development in Professionals. *Int. J. Sport. Med.* **2009**, *30*, 585–591. [CrossRef]
51. Kannekens, R.; Elferink-Gemser, M.T.; Visscher, C. Positioning and deciding: Key factors for talent development in soccer. *Scand. J. Med. Sci. Sport.* **2011**, *21*, 846–852. [CrossRef] [PubMed]
52. Teoldo da Costa, I.; de Freitas Silvino, M.P. Analysis of tactical behavior in full- and small-sided games: Comparing professional and youth academy athletes to enhance player development in soccer. *Int. J. Sport. Sci. Coach.* **2023**, *18*, 132–142. [CrossRef]
53. Sanchez, M.; Ramirez-Campillo, R.; Hernandez, D.; Carretero, M.; Luis-Pereira, J.M.; Sanchez-Sanchez, J. External Loads in Under-12 Players during Soccer-7, Soccer-8, and Soccer-11 Official Matches. *Int. J. Environ. Res. Public Health* **2021**, *18*, 4581. [CrossRef] [PubMed]
54. Bravo-Sánchez, A.; Abián-Vicén, J.; Abián, P. Analysis of the physical and technical differences between 7-a-side and 8-a-side game modalities in official under 12 soccer matches. *Int. J. Perform. Anal. Sport* **2017**, *17*, 545–554. [CrossRef]
55. Travassos, B.; Coutinho, D.; Gonçalves, B.; Pedroso, P.; Sampaio, J. Effects of manipulating the number of targets in U9, U11, U15 and U17 futsal players' tactical behaviour. *Hum. Mov. Sci.* **2018**, *61*, 19–26. [CrossRef] [PubMed]
56. Silva, A.F.; Conte, D.; Clemente, F.M. Decision-Making in Youth Team-Sports Players: A Systematic Review. *Int. J. Environ. Res. Public Health* **2020**, *17*, 3803. [CrossRef] [PubMed]
57. Williams, A.M. Perceptual skill in soccer: Implications for talent identification and development. *J. Sport. Sci.* **2000**, *18*, 737–750. [CrossRef] [PubMed]

58. Coutinho, D.; Gonçalves, B.; Wong, D.P.; Travassos, B.; Coutts, A.J.; Sampaio, J. Exploring the effects of mental and muscular fatigue in soccer players' performance. *Hum. Mov. Sci.* **2018**, *58*, 287–296. [CrossRef]
59. Davids, K.; Shuttleworth, R.; Button, C.; Renshaw, I.; Glazier, P. "Essential noise"—Enhancing variability of informational constraints benefits movement control: A comment on Waddington and Adams (2003). *Br. J. Sports Med.* **2004**, *38*, 601–605. [CrossRef]

sensors

MDPI

Article

Effect of Data-Processing Methods on Acceleration Summary Metrics of GNSS Devices in Elite Australian Football

Susanne Ellens, David L. Carey, Paul B. Gastin and Matthew C. Varley *

Sport, Performance, and Nutrition Research Group, School of Allied Health, Human Services and Sport, La Trobe University, Melbourne 3000, Australia; s.ellens@latrobe.edu.au (S.E.); d.carey@latrobe.edu.au (D.L.C.); p.gastin@latrobe.edu.au (P.B.G.)

* Correspondence: m.varley@latrobe.edu.au

Abstract: This study aimed to measure the differences in commonly used summary acceleration metrics during elite Australian football games under three different data processing protocols (raw, custom-processed, manufacturer-processed). Estimates of distance, speed and acceleration were collected with a 10-Hz GNSS tracking technology device from fourteen matches of 38 elite Australian football players from one team. Raw and manufacturer-processed data were exported from respective proprietary software and two common summary acceleration metrics (number of efforts and distance within medium/high-intensity zone) were calculated for the three processing methods. To estimate the effect of the three different data processing methods on the summary metrics, linear mixed models were used. The main findings demonstrated that there were substantial differences between the three processing methods; the manufacturer-processed acceleration data had the lowest reported distance (up to 184 times lower) and efforts (up to 89 times lower), followed by the custom-processed distance (up to 3.3 times lower) and efforts (up to 4.3 times lower), where raw data had the highest reported distance and efforts. The results indicated that different processing methods changed the metric output and in turn alters the quantification of the demands of a sport (volume, intensity and frequency of the metrics). Coaches, practitioners and researchers need to understand that various processing methods alter the summary metrics of acceleration data. By being informed about how these metrics are affected by processing methods, they can better interpret the data available and effectively tailor their training programs to match the demands of competition.

Keywords: data processing; smoothing; filter; GPS; acceleration

Citation: Ellens, S.; Carey, D.L.; Gastin, P.B.; Varley, M.C. Effect of Data-Processing Methods on Acceleration Summary Metrics of GNSS Devices in Elite Australian Football. *Sensors* **2024**, *24*, 4383. https://doi.org/10.3390/s24134383

Academic Editors: Bruno Gonçalves, Bruno Emanuel Nogueira Figueira and Diogo Alexandre Martins Coutinho

Received: 14 May 2024
Revised: 28 June 2024
Accepted: 2 July 2024
Published: 5 July 2024

1. Introduction

Global navigation satellite systems (GNSS) are a commonly used athlete tracking system in team sports and permit the quantification of player movement [1,2]. A GNSS device accesses satellites from multiple constellations in orbit (e.g., GPS and GLONASS) to determine its position in space, allowing the estimation of its distance covered, speed and acceleration [3,4]. In addition, some athlete tracking systems also include a triaxial accelerometer, gyroscope and magnetometer, allowing human activity recognition and the measurement of variables such as PlayerLoad™ [5,6]. The accelerometers within the athlete tracking system are not involved in the calculation of GNSS acceleration; accelerometer-derived acceleration is distinctly different data. Notably, most researchers and practitioners (79%) in team sports use GNSS-derived acceleration data [7]. Accurately quantifying player movements by determining the intensity, frequency and volume of these movements demonstrates the demands of a sport [8]. This knowledge can be used by practitioners to design training programs that adequately prepare athletes for competition [8].

Acceleration-based movements have been highlighted across the literature as important for team sport performance [9], as many movements require an athlete to accelerate or decelerate (negative acceleration) rapidly. Within sports, GNSS time-series acceleration data are often summarised according to the distance run or number of efforts performed within certain acceleration/deceleration bands [1,7,9]. Acceleration efforts have been identified as a critical component of Australian rules football (AF), where acceleration match demands increase with increasing competition level [10]. Therefore, quantifying these acceleration-based movements in AF are of interest to coaches and practitioners. Manufacturer data processing can have a large influence on GNSS acceleration data and corresponding summary metrics such as number of efforts [11,12]. Different data processes can alter the quantification of player movement, which may affect a practitioner's interpretation of the data and training programs. For example, applying a data processing method with strong smoothing, can cause a reduction in the number of acceleration efforts recorded during a match [11], potentially changing a practitioners interpretation of players' workload. Valid GNSS acceleration data are needed to correctly quantify player movement with summary metrics.

Various validation studies have assessed the ability of GNSS to estimate acceleration [12–16]. This is an ongoing process as each new GNSS device requires validation [17]. Numerous GNSS manufacturers apply their own data processing methods within their software, which could impact data validity and result in manufacturer-influenced variations in summary metrics, rather than being directly related to what the GNSS device is measuring [17]. Within the literature, large manufacturer-influenced variations have been reported in summary acceleration metrics [11,18]. Variations of up to ~250 acceleration efforts have been observed when using different manufacturer data processing methods on an identical dataset from a soccer match [11]. Although it is known that differences exist between manufacturer software processed data, most practitioners still use manufacturer's software-derived GNSS data, as it is a simple and efficient way to obtain data.

Practitioners and researchers experienced with data processing techniques might choose to extract and process the raw (not smoothed in any way by the manufacturer software) data from the GNSS devices and analyse it separately [9,19]. This approach offers several advantages such as eliminating undesired processing practices (e.g., smoothing and algorithms) and incorporating custom processes such as new summary metrics [9,17]. Custom processing of manufacturer-exported GNSS data has been shown to enhance acceleration data quality, and derived summary metrics differed compared to manufacturer-processed data [20]. When using custom processing on manufacturer-processed GNSS data, double processing and over-smoothing of the data could take place, which could eliminate important parts of the acceleration data. Using custom processing on raw GNSS data would eliminate double processing and could enhance data quality. However, there is no research comparing custom-processed raw GNSS data to manufacturer-processed data.

While the three methods (raw, custom-processed and manufacturer-processed) are available for analysis of GNSS data, there is no research identifying differences in the summary acceleration metrics between the methods. Furthermore, there is limited research exploring differences between summary acceleration metrics of just the raw and manufacturer-processed data. Large differences have been reported in the distance covered when accelerating between raw and manufacturer-processed data in a controlled environment where GNSS devices were positioned on a sled [18], and large differences have been found between just the acceleration data of team sport training sessions [20]. However, no study has investigated the difference in commonly used acceleration metrics between raw and manufacturer-processed data of team sport players during competition match-play.

Therefore, the aim of this study was to compare and explore the differences between three data processing methods (I. raw; II. custom-processed; III. manufacturer-processed) in the commonly used acceleration metrics of elite Australian football competition match-play data using GNSS tracking technology. To make this research practical, it was decided that GNSS-derived acceleration data would be used rather than acceleration data measured by an accelerometer, as most researchers and practitioners in team sports use GNSS-derived acceleration data.

2. Methods

2.1. Participants

Player movement data from thirty-eight elite male players from one Australian Football League team were collected with a 10-Hz GNSS tracking technology device from fourteen matches during the 2019 competitive season. Data were included if the horizontal dilution of precision (HDOP) was $\leq$2 and the player was in play on the field. There was no minimum playing duration requirement for a player file to be included. This resulted in a total sample size of 262 player files. The procedures used in this study were conducted with approval from the Human Research Ethics Committee of La Trobe University (reference number: HEC21282).

2.2. Equipment

Player movement, including estimations of acceleration, speed and distance, was measured using a 10-Hz GNSS device (Vector S7, Catapult Innovations, Melbourne, Australia). The device was positioned between the athlete's shoulder blades using the manufacturer's snug-fitting garment to prevent unnecessary device movement. Data collection procedures adhered to the guidelines outlined by Malone, Lovell, Varley and Coutts [17], with each athlete having their own specific device. The study sample had an average ($\pm$SD), 12 $\pm$ 1 number of satellites and horizontal dilution of precision (HDOP) of 0.62 $\pm$ 0.07.

The Vector S7 has reported acceptable levels of reliability and validity for speed (coefficient of variation $\leq$2%, mean bias $-$0.5%) and reliability and validity for distance (coefficient of variation $\leq$1.3%, mean bias $\leq$1%) according to Catapult's vector data integrity testing [21] and peer reviewed research [22].

The raw (not smoothed in any way by the manufacturer software) GNSS Doppler-shift speed data were exported and retrieved from the Catapult software (Openfield, version 2.7.1, Catapult Sports, Melbourne, Australia) files folder. The raw acceleration dataset was calculated using a central difference method on the raw Doppler-shift speed data. To determine the most appropriate custom processing method, several common smoothing methods (Butterworth filter: cut-off frequencies 0.1 to 4.9 Hz, exponential smoothing: smoothing constant 0.1 to 0.9, moving average: sliding window 0.1 s to 0.9 s) have been applied to the raw GNSS Doppler-shift speed data and were compared with a gold standard motion analysis system (Vicon) dataset. The fourth order (zero lag) low-pass Butterworth filter with a cut-off frequency of 2 Hz showed the strongest relationship with the Vicon data (mean bias 0.00 $m \cdot s^{-2}$, 95% LoA $\pm$ 1.55 $m \cdot s^{-2}$, RMSE 0.79 $m \cdot s^{-2}$) and was therefore used on the raw GNSS Doppler-shift speed data for the custom processing method. After using the Butterworth filter, acceleration was calculated using a central difference method on the custom-processed GNSS Doppler-shift speed data. Applying the processing to the raw Doppler-shift speed data before deriving acceleration data will ensure that any noise present in the raw Doppler-shift speed data will not be increased due to deriving acceleration. The manufacturer-processed GNSS distance and acceleration data were exported from the manufacturer's software using their default settings (Openfield, version 2.7.1, Catapult Sports, Melbourne, Australia). A summary of the details of the three datasets used for further analysis, (I) raw, (II) custom-processed, (III) manufacturer-processed, can be found in Table 1.

Table 1. Summary of the three used data processing methods and details on how acceleration data were obtained for each processing method.

Data Processing Method	How the Acceleration Data Were Obtained
Raw	Central difference method applied to raw GNSS Doppler-shift speed data to calculate acceleration.
Custom	Raw GNSS Doppler-shift speed data were processed with a fourth order (zero lag) low-pass Butterworth filter with a cut-off frequency of 2 Hz, whereafter acceleration was calculated using a central difference method on the processed GNSS Doppler-shift speed data.
Manufacturer	GNSS acceleration data were directly exported from manufacturer software using their default settings.

As the custom-processed data were derived from the raw data, these datasets were automatically synchronised. To allow for comparison of all results, the manufacturer-processed data were synchronised with the raw data. The raw data files represented all data from the time the GNSS units were switched on to start data collection until they were switched off. However, the manufacturer-processed data represented only gametime. Consequently, the files varied in length and could not be synchronised by means of cross-correlation. Unix timestamps (also known as epoch time) were used for synchronisation. The raw data files only included one Unix timestamp corresponding to the time the GNSS units were switched on to start data collection; they did not include a timestamp variable. The Unix timestamp was used to create a timeseries for the raw datafiles by extending the Unix timestamp by the sampling frequency of the device and the length of each file. The Unix timeseries of the raw and manufacturer-processed datasets were then used to synchronise and join both datasets. Cross-correlation analysis was performed afterwards to confirm perfect alignment of the raw and manufacturer-processed data.

2.3. Data Analysis

Two common summary acceleration metrics were extracted from the datasets. The first metric was the number of high- and medium-intensity acceleration and deceleration efforts which were extracted for each player and game from each of the three acceleration datasets. The start of a high effort was defined by a $\pm 3\ m \cdot s^{-2}$ threshold (a negative threshold defines a deceleration effort and a positive threshold an acceleration effort) and $\pm 2\ m \cdot s^{-2}$ for medium efforts. These thresholds were selected as they are commonly used for high and medium acceleration and deceleration efforts in the research literature [7,9,23]. An effort was counted when the acceleration data reached the set threshold and stayed above the set threshold for at least 0.3 s [24] and ended when the acceleration data reached $0\ m \cdot s^{-2}$ [11]. The second metric was the distance covered in meters using the manufacturer-processed GNSS distance within a predefined high ($\geq 3\ m \cdot s^{-2}$ for acceleration and $\leq -3\ m \cdot s^{-2}$ for deceleration) and medium (2 to $3\ m \cdot s^{-2}$ for acceleration and -2 to $-3\ m \cdot s^{-2}$ for deceleration) intensity zone, extracted from each of the three acceleration datasets.

2.4. Statistical Analysis

To estimate the effect of the three different data processing methods on the number of acceleration/deceleration efforts (#) and distance (m) within the high-intensity zone ($\pm 3\ m \cdot s^{-2}$) or medium-intensity zone ($\pm 2\ m \cdot s^{-2}$), linear mixed models were used to account for recurring measures. A negative binomial generalised linear mixed model was used for the effort model, as the efforts were count-based, not normally distributed data [25], and a linear mixed model was used for the distance model [26]. Each model included a fixed effect for processing method (raw, custom-processed, manufacturer-processed). The models included a random effect for player ID and game, which allowed for different mean values for each player and game. The change in number of efforts or distance reported between processing methods within the medium- or high-acceleration or -deceleration zone was estimated, and a 95% confidence interval (CI) was used to denote the imprecision of the fixed effect parameter estimates. To determine the difference in number of efforts (#) and distance (m) within each processing method between the high-intensity zone or medium-intensity zone, the same mixed models were used as mentioned above, but with intensity (medium or high) as a fixed effect. All analysis were performed in MATLAB (version 9.14.0 (R2023a), The MathWorks Inc., Natick, MA, USA).

3. Results

3.1. Between Processing Methods Effects

Overall, manufacturer processing had the lowest reported distance and efforts, followed by the custom processing, then the raw data. When using manufacturer-processed data, distance covered while accelerating above the high threshold was on average 7.5 m, whereas the custom processing brought the distance up to an average of 421 m and the raw data, 1380 m (Figure 1). For the efforts reported in the high-acceleration zone, manufacturer processing reported 3 efforts on average, where custom processing reported 138 efforts and raw data 221 efforts.

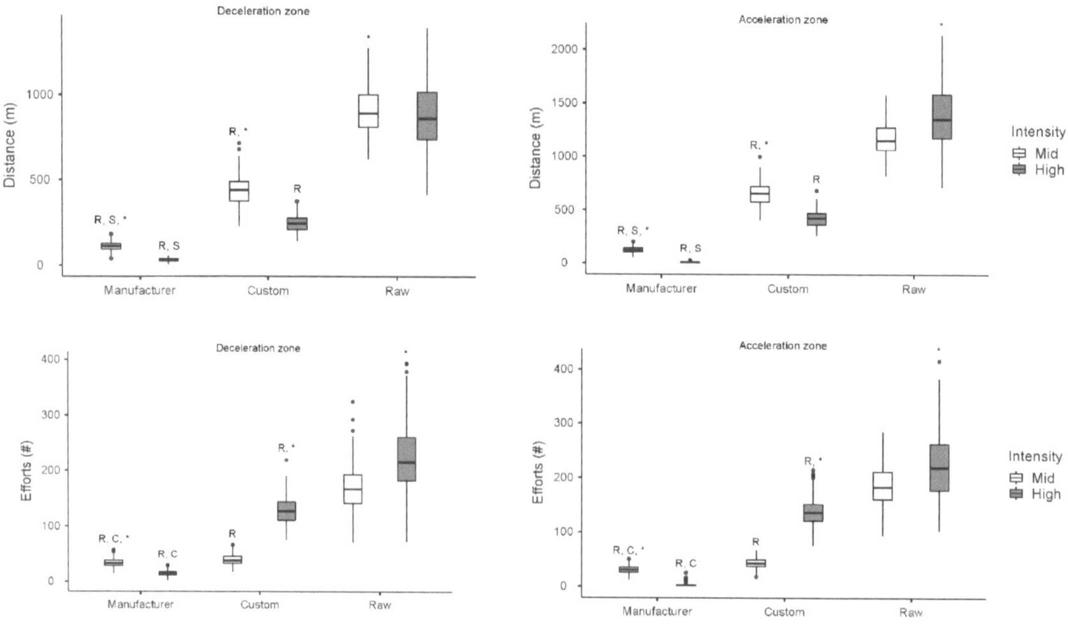

Figure 1. Distance (m) decelerating (**left–top**) and accelerating (**right–top**), number of deceleration efforts (**left–bottom**) and acceleration efforts (**right–bottom**), calculated for Medium (± 2 m·s^{-2}) and High (± 3 m·s^{-2}) zones for one dataset processed in three different ways (I) raw, (II) custom-processed, (III) manufacturer-processed. R = Data are different compared to Raw dataset of the same intensity. C = Data are different compared to custom-processed dataset of the same intensity. * = Significantly higher distance compared to the other intensity which was processed the same way.

For the variable distance in meters while in the high deceleration zone, the main effect for processing method was significant ($F(2, 696) = 5371$, $p < 0.001$). Comparing the custom processing to manufacturer processing, distance increased by 217 m, (95% confidence interval (CI) = [200 m to 233 m], $t(696) = 25$, $p < 0.001$). Comparing the raw data to manufacturer processing, distance increased by 849 m, (95% CI = [832 m to 865 m], $t(696) = 100$, $p < 0.001$). Comparing the raw data to custom processing, distance increased by 632 m, (95% CI = [615 m to 649 m], $t(696) = 74$, $p < 0.001$). The results of the effect of processing method on distance in the medium/high-acceleration or -deceleration zones are presented in Table 2.

Table 2. Effects of processing method (raw, custom-processed, manufacturer-processed), on distance in meters split by medium (±2 m·s^{-2})- and high (±3 m·s^{-2})-acceleration or -deceleration intensity.

Effect of Processing Method on Distance								
	Intensity	Effect	Estimate (m)	Lower 95% CI	Higher 95% CI	df	t	p
Acceleration	High	Custom–Manufacturer	413	389	437	697	34	<0.001
		Raw–Manufacturer	1373	1349	1397	697	112	<0.001
		Raw–Custom	959	935	983	696	78	<0.001
	Medium	Custom–Manufacturer	529	515	543	719	74	<0.001
		Raw–Manufacturer	1042	1028	1056	719	145	<0.001
		Raw–Custom	513	499	527	719	71	<0.001
Deceleration	High	Custom–Manufacturer	217	200	233	696	25	<0.001
		Raw–Manufacturer	849	832	865	696	100	<0.001
		Raw–Custom	632	615	649	696	74	<0.001
	Medium	Custom–Manufacturer	327	315	339	719	53	<0.001
		Raw–Manufacturer	798	786	810	719	130	<0.001
		Raw–Custom	471	459	484	719	76	<0.001

df = degrees of freedom; t = t-statistic; *p* = *p*-value.

For the variable number of efforts while in the high-deceleration zone, the main effect for processing method was significant, $X^2(2) = 15{,}938$, $p < 0.001$. Comparing the custom processing to manufacturer processing, the number of efforts increased 8.88 times, (95% CI = [8.51 to 9.57], $p < 0.001$). Comparing the raw data to manufacturer processing, the number of efforts increased 15.2 times, (95% CI = [14.6 to 15.9], $p < 0.001$). Comparing the raw data to custom processing, the number of efforts increased 1.71 times, (95% CI = [1.67 to 1.77], $p < 0.001$). The results of the effect of processing method on number of efforts in the medium/high acceleration or deceleration zones are presented in Table 3.

Table 3. Effects of processing method (raw, custom-processed, manufacturer-processed), on number of efforts split by medium (±2 m·s^{-2}) and high (±3 m·s^{-2}) acceleration or deceleration intensity.

Effect of Processing Method on Number of Efforts						
	Intensity	Effect	Estimate (Rate of Change)	Lower 95% CI	Higher 95% CI	p
Acceleration	High	Custom–Manufacturer	55.7	51.3	60.4	<0.001
		Raw–Manufacturer	89.7	82.6	97.4	<0.001
		Raw–Custom	1.61	1.57	1.66	<0.001
	Medium	Custom–Manufacturer	1.43	1.38	1.48	<0.001
		Raw–Manufacturer	6.13	5.93	6.34	<0.001
		Raw–Custom	4.29	4.16	4.43	<0.001
Deceleration	High	Custom–Manufacturer	8.88	8.51	9.57	<0.001
		Raw–Manufacturer	15.2	14.6	15.9	<0.001
		Raw–Custom	1.71	1.67	1.77	<0.001
	Medium	Custom–Manufacturer	1.18	1.14	1.23	<0.001
		Raw–Manufacturer	5.09	4.92	5.27	<0.001
		Raw–Custom	4.30	4.16	4.45	<0.001

3.2. Within Processing Method Effects

Overall, the distance while accelerating or decelerating was largest in the medium zone compared to the high zone for all processing methods, except for distance while accelerating for the raw method. The number of efforts was largest in the high zone compared to the medium zone for all processing methods except manufacturer processing.

For the variable distance while decelerating processed by the manufacturer, the main effect for intensity was significant (F(1, 453) = 4345, $p < 0.001$). Distance in the medium-intensity zone increased by 79 m compared to the high-intensity zone (95% CI = [77 m to 81 m], t(453) = 66, $p < 0.001$). The results of the effect of intensity on distance while decelerating or accelerating for each processing method are presented in Table 4.

Table 4. Effects of acceleration or deceleration intensity, medium ($\pm 2\ m \cdot s^{-2}$) and high ($\pm 3\ m \cdot s^{-2}$), on distance in meters split by processing method (raw, custom-processed, manufacturer-processed).

Effect of Intensity on Distance								
	Processing	**Effect**	**Estimate (m)**	**Lower 95% CI**	**Higher 95% CI**	**df**	**t**	***p***
Acceleration	Manufacturer	Medium–High	118	114	121	455	74	<0.001
	Custom	Medium–High	234	225	243	455	51	<0.001
	Raw	Medium–High	−211	−234	−189	455	−18	<0.001
Deceleration	Manufacturer	Medium–High	79	77	81	453	66	<0.001
	Custom	Medium–High	190	182	197	455	52	<0.001
	Raw	Medium–High	29	13	45	455	3.4	<0.001

df = degrees of freedom; t = t-statistic; *p* = *p*-value.

For the variable number of efforts while decelerating processed by the manufacturer, the main effect for intensity was significant ($X^2(2) = 1368$, $p < 0.001$). The number of efforts in the medium-intensity zone was 2.28 times greater compared to the high-intensity zone (95% CI = [2.18 to 2.38], $p < 0.001$). The results of the effect of intensity on number of efforts while decelerating or accelerating for each processing method are presented in Table 5.

Table 5. Effects of acceleration or deceleration intensity, medium ($\pm 2\ m \cdot s^{-2}$) and high ($\pm 3\ m \cdot s^{-2}$) on number of efforts, split by processing method (raw, custom-processed, manufacturer-processed).

Effect of Intensity on Number of Efforts						
	Processing	**Effect**	**Estimate (Rate of Change)**	**Lower 95% CI**	**Higher 95% CI**	***p***
Acceleration	Manufacturer	Medium–High	12.1	11.1	13.1	<0.001
	Custom	Medium–High	0.71	0.63	0.79	<0.001
	Raw	Medium–High	0.83	0.81	0.85	<0.001
Deceleration	Manufacturer	Medium–High	2.28	2.18	2.38	<0.001
	Custom	Medium–High	0.30	0.29	0.31	<0.001
	Raw	Medium–High	0.66	0.60	0.72	<0.001

4. Discussion

This study aimed to measure the differences in commonly used summary acceleration metrics during elite Australian football games of GNSS acceleration data that were derived using three different processing methods (raw, custom-processed, manufacturer-processed). The main finding was that there were substantial differences between the three processing methods when calculating the same metric. Overall, compared to the raw data, the manufacturer-processed acceleration data had the lowest reported distance (up to 184 times lower) and efforts (up to 89 times lower), followed by the custom-processed distance (up to 3.3 times lower) and efforts (up to 4.3 times lower), where raw data had the highest reported distance and efforts.

The raw data were unprocessed and consequently had the most noise present, resulting in the highest distance covered and number of efforts. The manufacturer-processed acceleration data had the lowest reported distance and efforts. The results were approximately 28 efforts lower than those found in literature using a similar GNSS device and manufacturer software [27]. The difference could be explained by the fact that Rennie, Kelly, Bush, Spurrs, Austin and Watsford [27] used a lower threshold of $\pm 2.78\ m{\cdot}s^{-2}$ and a shorter duration above the set threshold of 0.2 s to identify an effort. A lower duration above the threshold of 0.2 s vs. 0.3 s (which was used in this study) has been shown to identify 45% more acceleration efforts and 13% more deceleration efforts [24]. Furthermore, all data that involved <75% of total game time were excluded from Rennie, Kelly, Bush, Spurrs, Austin and Watsford [27], while there was no minimum game time requirements for this study. When taking all factors (lower threshold, shorter duration above the threshold and game time criteria) into consideration, the number of efforts reported are comparable to the current study. This further highlights the effects of different processing methods on acceleration data and accompanying difficulty in comparing results across studies.

Coaches and practitioners use acceleration metrics to quantify the demands (volume, intensity and frequency of the metrics) of a sport or activity [7,28], which can be used to create training programs to adequately prepare players for competition [8]. Furthermore, researchers could be using and analysing GNSS manufacturer-processed acceleration data and metrics for their research. A change in the metric output due to processing methods will alter the demands they are quantifying. For example, a sudden increase in the volume of acceleration undertaken might indicate to a coach that players are working harder than normal. Although they are undertaking the same amount of work as usual, the metric output increased due to different processing methods used. Examples of when processing methods might change include when software is updated, athlete tracking technologies are changed or when different software/algorithms are used to process the data. Coaches, practitioners and researchers should be aware that processing methods can change, and that these changes could affect metric outputs and alter the demands they are quantifying.

To be able to select a suitable processing technique for acceleration data, one should be aware of the characteristics of their data (patterns and frequencies that could be present in the data). The characteristics of the data can determine what type of processing method is most suitable [29]. Based on the results of this study, practitioners using GNSS acceleration data are recommended to select a processing method specific to their use case and characteristics of the data. It is also recommended to evaluate the impact of different processing methods on metrics of interest (e.g., number of efforts). For this study, the characteristics of team sport-specific human movement patterns (elite AF players) and potential sources of variability in the acceleration data should be taken into consideration. An athlete in full sprint could have anywhere between 2 and 5 steps per second [30], indicating that at least 2 Hz patterns (corresponding to 2 steps per second) could be present in the acceleration data. The acceleration data varies within a single step (from the heel strike of one foot to the subsequent heel strike of the other foot), which is a result of changing the balance between the propulsive and braking forces at each ground contact [31]. A surplus in propulsive forces results in acceleration, where a surplus of braking forces results in deceleration. If an athlete performs 2 steps per second, that means the acceleration data change significantly at each single step [32]. This suggests that acceleration data might be more variable and higher than what is currently indicated by GNSS devices.

The high acceleration values (considered as acceleration and deceleration values with a high rate of change in speed) are shown by the results of the custom processing method, which was smoothed with a filter which had a 2 Hz cutoff frequency, allowing for 2 Hz patterns in the data. This method might be the closest approximation to the real world of all three processing methods. The larger distance run and number of efforts in raw and custom-processed data compared to manufacturer-processed data is an indicator of stronger data smoothing in the manufacturer-processed data. The custom-processed and raw dataset showed that the acceleration data exceeded the high threshold (± 3 m·s^{-2}) for more than 0.3 s, significantly more (up to 89.7 times) than the manufacturer-processed dataset (see Table 3). Strong smoothing in the manufacturer data could eliminate important portions of a signal by smoothing peaks and lowering the amplitude of high acceleration data. The amplitude of the manufacturer-processed acceleration data is lowered to the point where more data exceed the medium threshold (± 2 m·s^{-2}) but stay below the high threshold (± 3 m·s^{-2}), as evident by the larger number of efforts in the medium zone compared to the high zone (see Table 5). When looking into the literature, athletes have reached acceleration values between 5–7 m·s^{-2} [32–35], suggesting that elite AF players should be able to reach these values. However, the manufacturer-processed data suggest that the elite AF players barely reach the set ± 3 m·s^{-2} threshold, which is an indication that manufacturer-processed data may be over-smoothing and masking the actual acceleration values that an athlete is capable of.

The application of a data processing method with strong smoothing has been shown to cause a reduction in the number of recorded acceleration efforts during a match [11]. Furthermore, manufacturer-processed GNSS acceleration data have shown a very large mean bias, with lower acceleration values, when compared to a criterion measure [12]. In combination with the results from the current study, these findings collectively suggest that manufacturer-processed data are subject to extensive data smoothing.

Distance run in the medium-acceleration zone exceeded that of the high zone for all different processing techniques except for the raw acceleration dataset, where the distance recorded in the high zone was greater. The raw acceleration dataset was not subject to smoothing, meaning that all potential sources of noise, such as sensor movement, multipath interference and environmental conditions [36], were present in the data. This noise may manifest as high-frequency components, leading to a larger representation of raw data in the high zone compared to the medium zone and all other processing methods. Similar findings have been reported for distance based summary acceleration metrics between raw and manufacturer-processed data in a controlled environment [18] and between the acceleration data of team sport training sessions [20].

Large inconsistencies exist in the literature for reported processing steps used on acceleration data, which hinders the comparison of acceleration summary metric results between studies [7,37]. The findings of the current study demonstrated substantial differences between different processing methods when estimating the same acceleration and deceleration metric. Therefore, future research should report all different processing steps performed on their used acceleration data derived from an athlete tracking system to ensure comparability of results between studies.

A noteworthy strength of this research is the use of elite Australian Football team data. The dataset consisted of games played on different days and at different locations within stadia, providing a real representation of diverse GNSS team data. The data were collected with one type of GNSS device and is thus only representative of this specific device. Furthermore, the manufacturer-processed data were exported using the specific manufacturer software mentioned in the methods section. Since data processing procedures could vary between software versions, it is important to note that the manufacturer-processed data are representative only of this specific software. Future research investigating processing methods of acceleration data of athlete tracking technologies should consider using local positioning systems (LPS) and optical positioning systems, next to GNSS. Current elite team sport environments require teams to use a variety of athlete tracking technologies suitable for different locations, e.g., LPS or optical for indoor stadia, GNSS for outdoor or training sessions [18]. The use of different athlete tracking systems interchangeably, requires research to establish the influence of data processing on acceleration data of different tracking systems to be able to compare acceleration data longitudinally.

5. Conclusions

The results from this study demonstrated that there were substantial differences in commonly used summary acceleration metrics (number of efforts performed and distance covered) during elite Australian football games between three processing methods (raw, custom-processed, manufacturer-processed). Overall, the manufacturer-processed acceleration data had the lowest reported distance and efforts, followed by the custom-processed distance, where raw data had the highest reported distance and efforts. The results indicate that different processing methods changed the metric output (number of efforts and distance covered) and can in turn alter the quantification of the demands of a sport (volume, intensity and frequency of the metrics). It is important for coaches, practitioners and researchers using GNSS-derived acceleration data to know how, and be aware that, processing methods change summary acceleration metrics (e.g., efforts and distance covered) because they are often used to quantify the demands of a sport and to create training programs to adequately prepare players for competition. Furthermore, it is recommended that future research and tracking technology manufacturers report all data processing practises performed on the acceleration data where possible. Knowing all performed processing steps allow for comparability of results and the ability to identify if differences may be due to processing practises rather than the used tracking technology.

Author Contributions: Conceptualization, S.E., D.L.C., P.B.G. and M.C.V.; methodology, S.E., D.L.C., P.B.G. and M.C.V.; software, S.E.; formal analysis, S.E.; investigation, S.E.; writing—original draft preparation, S.E.; writing—review and editing, D.L.C., P.B.G. and M.C.V.; visualization, S.E.; supervision, D.L.C., P.B.G. and M.C.V.; project administration, S.E.; All authors have read and agreed to the published version of the manuscript.

Funding: This research received no external funding.

Institutional Review Board Statement: All subjects gave their informed consent for inclusion before they participated in the study. The study was conducted in accordance with the Declaration of Helsinki, and the protocol was approved by the Ethics Committee of La Trobe University (reference number: HEC21282).

Informed Consent Statement: Informed consent was obtained from all subjects involved in the study.

Data Availability Statement: The data presented in this study are available on request from the corresponding author. The data are not publicly available due to ethical restrictions.

Conflicts of Interest: The authors report no conflicts of interest.

References

1. Cummins, C.; Orr, R.; O'Connor, H.; West, C. Global positioning systems (GPS) and microtechnology sensors in team sports: A systematic review. *Sports Med.* **2013**, *43*, 1025–1042. [CrossRef] [PubMed]
2. Torres-Ronda, L.; Clubb, J.; Beanland, E. Tracking systems in team sports: Back to basics. *Sport Perform. Sci. Rep.* **2022**, *159*, 1–9.
3. Silva, H.; Nakamura, F.Y.; Ribeiro, J.; Asian-Clemente, J.; Roriz, P.; Marcelino, R. Using minimum effort duration can compromise the analysis of acceleration and deceleration demands in football. *Int. J. Perform. Anal. Sport* **2023**, *23*, 125–137. [CrossRef]
4. Duthie, G.M.; Robertson, S.; Thornton, H.R. A GNSS-based method to define athlete manoeuvrability in field-based team sports. *PLoS ONE* **2021**, *16*, e0260363. [CrossRef] [PubMed]
5. Khan, D.; Al Mudawi, N.; Abdelhaq, M.; Alazeb, A.; Alotaibi, S.S.; Algarni, A.; Jalal, A. A Wearable Inertial Sensor Approach for Locomotion and Localization Recognition on Physical Activity. *Sensors* **2024**, *24*, 735. [CrossRef] [PubMed]
6. Gastin, P.B.; McLean, O.; Spittle, M.; Breed, R.V. Quantification of tackling demands in professional Australian football using integrated wearable athlete tracking technology. *J. Sci. Med. Sport* **2013**, *16*, 589–593. [CrossRef] [PubMed]
7. Ellens, S.; Middleton, K.; Gastin, P.B.; Varley, M.C. Techniques to derive and clean acceleration and deceleration data of athlete tracking technologies in team sports: A scoping review. *J. Sports Sci.* **2022**, *40*, 1772–1800. [CrossRef] [PubMed]
8. Akenhead, R.; Harley, J.A.; Tweddle, S.P. Examining the external training load of an English Premier League football team with special reference to acceleration. *J. Strength Cond. Res.* **2016**, *30*, 2424–2432. [CrossRef] [PubMed]
9. Delaney, J.A.; Cummins, C.J.; Thornton, H.R.; Duthie, G.M. Importance, reliability, and usefulness of acceleration measures in team sports. *J. Strength Cond. Res.* **2018**, *32*, 3485–3493. [CrossRef]
10. Johnston, R.D.; Black, G.M.; Harrison, P.W.; Murray, N.B.; Austin, D.J. Applied Sport Science of Australian Football: A Systematic Review. *Sports Med.* **2018**, *48*, 1673–1694. [CrossRef] [PubMed]

11. Varley, M.C.; Jaspers, A.; Helsen, W.F.; Malone, J.J. Methodological considerations when quantifying high-intensity efforts in team sport using global positioning system technology. *Int. J. Sports Physiol. Perform.* **2017**, *12*, 1059–1068. [CrossRef] [PubMed]
12. Delaney, J.A.; Wileman, T.M.; Perry, N.J.; Thornton, H.R.; Moresi, M.P.; Duthie, G.M. The validity of a global navigation satellite system for quantifying small-area team-sport movements. *J. Strength Cond. Res.* **2019**, *33*, 1463–1466. [CrossRef] [PubMed]
13. Scott, M.T.; Scott, T.J.; Kelly, V.G. The Validity and Reliability of Global Positioning Systems in Team Sport: A Brief Review. *J. Strength Cond. Res.* **2016**, *30*, 1470–1490. [CrossRef] [PubMed]
14. Brosnan, R.J.; Watson, G.; Stuart, W.; Twentyman, C.; Kitic, C.M.; Schmidt, M. The validity, reliability, and agreement of global positioning system units-can we compare research and applied data? *J. Strength Cond. Res.* **2022**, *36*, 3330–3338. [CrossRef] [PubMed]
15. Johnston, R.D.; Hewitt, A.; Duthie, G. Validity of Real-Time Ultra-wideband Global Navigation Satellite System Data Generated by a Wearable Microtechnology Unit. *J. Strength Cond. Res.* **2019**, *34*, 2071–2075. [CrossRef]
16. Linke, D.; Link, D.; Lames, M. Validation of electronic performance and tracking systems EPTS under field conditions. *PLoS ONE* **2018**, *13*, e0199519. [CrossRef] [PubMed]
17. Malone, J.J.; Lovell, R.; Varley, M.C.; Coutts, A.J. Unpacking the black box: Applications and considerations for using gps devices in sport. *Int. J. Sport Physiol. Perform.* **2017**, *12*, 18–26. [CrossRef] [PubMed]
18. Thornton, H.R.; Nelson, A.R.; Delaney, J.A.; Serpiello, F.R.; Duthie, G.M. Interunit Reliability and Effect of Data-Processing Methods of Global Positioning Systems. *Int. J. Sport Physiol. Perform.* **2019**, *14*, 432–438. [CrossRef] [PubMed]
19. Duthie, G.; Thornton, H.; Delaney, J.; McMahon, J.; Benton, D. Relationship between physical performance testing results and peak running intensity during professional rugby league match play. *J. Strength Cond. Res.* **2020**, *34*, 3506–3513. [CrossRef]
20. Delves, R.I.M.; Duthie, G.M.; Ball, K.A.; Aughey, R.J. Applying common filtering processes to Global Navigation Satellite System-derived acceleration during team sport locomotion. *J. Sports Sci.* **2022**, *40*, 1116–1126. [CrossRef] [PubMed]
21. Catapult. *Vector Data Validation*; Catapult: Melbourne, Australia, 2020.
22. Crang, Z.L.; Duthie, G.; Cole, M.H.; Weakley, J.; Hewitt, A.; Johnston, R.D. The inter-device reliability of global navigation satellite systems during team sport movement across multiple days. *J. Sci. Med. Sport* **2021**, *25*, 340–344. [CrossRef]
23. Oliva-Lozano, J.M.; Fortes, V.; Krustrup, P.; Muyor, J.M. Acceleration and sprint profiles of professional male football players in relation to playing position. *PLoS ONE* **2020**, *15*, e0236959. [CrossRef] [PubMed]
24. Ellens, S.; Carey, D.; Gastin, P.; Varley, M.C. Changing the criteria applied to acceleration and deceleration efforts changes the types of player actions detected. *Sci. Med. Footb.* **2022**, *8*, 52–59. [CrossRef] [PubMed]
25. McCullagh, P.; Nelder, J.A. *Generalized Linear Models*, 2nd ed.; Chapman and Hall: London, UK, 1989.
26. Newans, T.; Bellinger, P.; Drovandi, C.; Buxton, S.; Minahan, C. The Utility of Mixed Models in Sport Science: A Call for Further Adoption in Longitudinal Data Sets. *Int. J. Sport Physiol. Perform.* **2022**, *17*, 1289–1295. [CrossRef] [PubMed]
27. Rennie, M.J.; Kelly, S.J.; Bush, S.; Spurrs, R.W.; Austin, D.J.; Watsford, M.L. Phases of match-play in professional Australian Football: Distribution of physical and technical performance. *J. Sports Sci.* **2020**, *38*, 1682–1689. [CrossRef]
28. Delves, R.I.M.; Aughey, R.J.; Ball, K.; Duthie, G.M. The quantification of acceleration events in elite team sport: A systematic review. *Sports Med.-Open* **2021**, *7*, 45. [CrossRef] [PubMed]
29. Winter, D.A. *Biomechanics and Motor Control of Human Movement*; John Wiley & Sons: Hoboken, NJ, USA, 2009.
30. Kawabata, M.; Goto, K.; Fukusaki, C.; Sasaki, K.; Hihara, E.; Mizushina, T.; Ishii, N. Acceleration patterns in the lower and upper trunk during running. *J. Sport Sci.* **2013**, *31*, 1841–1853. [CrossRef] [PubMed]
31. Haugen, T.; McGhie, D.; Ettema, G. Sprint running: From fundamental mechanics to practice-a review. *Eur. J. Appl. Physiol.* **2019**, *119*, 1273–1287. [CrossRef] [PubMed]
32. Nagahara, R.; Kanehisa, H.; Fukunaga, T. Ground reaction force across the transition during sprint acceleration. *Scand. J. Med. Sci. Sports* **2020**, *30*, 450–461. [CrossRef] [PubMed]
33. Morin, J.B.; Le Mat, Y.; Osgnach, C.; Barnabo, A.; Pilati, A.; Samozino, P.; di Prampero, P.E. Individual acceleration-speed profile in-situ: A proof of concept in professional football players. *J. Biomech.* **2021**, *123*, 110524. [CrossRef]
34. Morin, J.D.; Bourdin, M.; Edouard, P.; Peyrot, N.; Samozino, P.; Lacour, J.R. Mechanical determinants of 100-m sprint running performance. *Eur. J. Appl. Physiol.* **2012**, *112*, 3921–3930. [CrossRef] [PubMed]
35. di Prampero, P.E.; Fusi, S.; Sepulcri, L.; Morin, J.B.; Belli, A.; Antonutto, G. Sprint running: A new energetic approach. *J. Exp. Biol.* **2005**, *208*, 2809–2816. [CrossRef] [PubMed]
36. Liu, Q.; Ying, R.; Dai, Z.; Wang, Y.; Qian, J.; Liu, P. Multi-Phase Fusion for Pedestrian Localization Using Mass-Market GNSS and MEMS Sensors. *Sensors* **2023**, *23*, 3624. [CrossRef] [PubMed]
37. Linke, D.; Link, D.; Lames, M. Football-specific validity of TRACAB's optical video tracking systems. *PLoS ONE* **2020**, *15*, e0230179. [CrossRef] [PubMed]

sensors

MDPI

Article

Do Elite Basketball Players Maintain Peak External Demands throughout the Entire Game?

Hugo Salazar [1], Filip Ujakovic [2], Jernej Plesa [2], Alberto Lorenzo [3,*] and Enrique Alonso-Pérez-Chao [4,5]

1 Faculty of Education and Sport, University of Basque Country (UPV/EHU), 01007 Vitoria-Gasteiz, Spain; hsalazar002@gmail.com
2 Basketball Club Cedevita Olimpija, 1000 Ljubljana, Slovenia; filip.ujakovic@student.kif.unizg.hr (F.U.)
3 Faculty of Physical Activity and Sport Sciences—INEF, Polytechnic University of Madrid, 28040 Madrid, Spain
4 Department of Physical Activity and Sports Science, University Alfonso X el Sabio, 28691 Villanueva de la Cañada, Spain; ealonsoperezchao@gmail.com
5 Faculty of Sports Science, European University of Madrid, 28670 Villaviciosa de Odón, Spain
* Correspondence: alberto.lorenzo@upm.es

Abstract: Consideration of workload intensity and peak demands across different periods of basketball games contributes to understanding the external physical requirements of elite basketball players. Therefore, the aim of this study was to investigate the average intensity and peak demands encountered by players throughout game quarters. PlayerLoad per minute and PlayerLoad at three different time samples (30 s, 1 min, and 3 min) were used as workload metrics. A total of 14 professional elite male basketball players were monitored during 30 official games to investigate this. A linear mixed model and Cohen's d were employed to identify significant differences and quantify the effect sizes among game quarters. The results showed a significant, moderate effect in PlayerLoad per minute between Q1 vs. Q4, and a small effect between Q2 and Q3 vs. Q4. Furthermore, a small to moderate decline was observed in external peak values for PlayerLoad across game quarters. Specifically,, a significant decrease was found for the 3 min time window between Q1 and other quarters. The findings from the present study suggest that professional basketball players tend to experience fatigue or reduced physical output as the game progresses.

Keywords: basketball; most demanding scenarios; PlayerLoad; team sport; physical demands; accelerometry

Citation: Salazar, H.; Ujakovic, F.; Plesa, J.; Lorenzo, A.; Alonso-Pérez-Chao, E. Do Elite Basketball Players Maintain Peak External Demands throughout the Entire Game? *Sensors* **2024**, *24*, 4318. https://doi.org/10.3390/s24134318

Academic Editors: Georg Fischer and Arnold Baca

Received: 2 May 2024
Revised: 29 May 2024
Accepted: 24 June 2024
Published: 3 July 2024

1. Introduction

Load monitoring has become an essential process for coaches and sports science practitioners to examine the individual workload of players and the collective workload of teams. Additionally, quantifying physical and physiological loads is important for understanding the dose–response nature of the training process when establishing optimal training procedures [1]. Training load includes both external and internal components. External loads (ELs) indicate the physical workload performed (e.g., duration, distance), which is determined by the organization, quality, and quantity of exercise (training plan) [2]. Internal load (IL) refers to the psycho-physiological response during exercise aimed at meeting the demands imposed by the EL (e.g., heart rate, heart rate variability, rate of perceived exertion) [2]. In the modern era of sports science, recent technological advancements, such as electronic performance tracking systems (EPTSs) integrating inertial measurement units (IMUs) like accelerometers and gyroscopes, have transformed the monitoring of basketball players in both training sessions and actual games [3].

Recent studies utilizing EPTSs have characterized basketball as an intermittent, high-intensity sport. These studies have shown that the majority of playing time (93.65%) is spent in standing–walking (<7 km/h) and jogging activities (7–14 km/h) [4]. These periods of low-motion activities are interspersed with the most demanding scenarios (e.g., peak demands, high-intensity period). During these moments, players constantly engage in continuous changes in direction, jumps, accelerations, decelerations, physical contacts, and specific basketball skills (e.g., crossovers, lay-ups) [5,6]. Investigating the physical demands experienced by athletes during both competition and training has emerged as a focal point in sports science research [7]. This expanding field enables sports coaches to gather precise data for refining athletic training programs to target specific adaptations. To accomplish this, it is crucial to implement training methods that not only capture average values but also replicate the challenges posed by peak competition demands, thereby optimizing overall athletic performance [6].

Peak demands (PDs) refer to the most intense activity experienced by players within a specified timeframe for a chosen variable [6]. PDs in basketball players have been assessed using various external variables (e.g., PlayerLoad or distance) and time windows (e.g., 30 s, 45 s, 1 min, 5 min, and 10 min) [6,8–10]. The literature on PDs in basketball has examined the influence of contextual factors such as player position [6,11], game score-line [12,13], game schedule [14], team venue [15], cumulative playing time across the entire game [9,16] and prior to intense passages [9], activity type [17,18], age category [19] or moment of the game [10,20]. However, these studies are often conducted on non-professional samples, primarily due to the challenges associated with accessing professional players.

Previous research on professional basketball has examined the impact of various factors on the external PDs experienced by players, including playing positions [21,22] and the seasonal period [23]. There is a dearth of studies in the literature that assess professional or elite basketball players, potentially due to the challenges associated with accessing data within high-performance environments or the limited availability of microtechnology for quantifying competition demands in certain professional leagues [24]. Furthermore, the imprecise usage of terms such as "elite", "high performance", and "professional" in sports contexts introduces ambiguity and may result in research spanning various competitive levels being classified as "elite". This lack of specificity complicates the interpretation and comparison of findings across studies [25,26].

Understanding the fluctuations in PDs provides a deeper insight into the demands of competition. Fluctuations refer to variations in the intensity, measurement, or quality of a variable over a period of time [20]. One of the earlier published articles evaluating fluctuations throughout the quarters of a game demonstrated how dribbling actions and total activity velocities declined as the game progressed [27]. Furthermore, with the implementation of microtechnology during games and among professional second-division players, it has been revealed that all average external physical demands decreased across game quarters [21]. Expanding the scope to include different levels of play, research conducted among semi-professional [10,22] or youth basketball players [20] has uncovered a similar trend in peak demands as observed in average demands, with higher peak values typically recorded in the first quarter compared to the latter stages of the game [10,20,22]. Collectively, these findings underscore the intricate dynamics of physical performance throughout a basketball game, influenced by factors such as player expertise, match intensity, and strategic maneuvers employed by the teams [10,20,22].

Based on the preceding information, our understanding of the behavior of external peak demand during official basketball matches among elite or professional players remains limited. Consequently, it cannot be assumed that the differences in external PDs observed between quarters can be generalized to other samples (e.g., female players or professional players) [9]. Furthermore, separate research on this topic is necessary for elite male players, as top male basketball players have been shown to face significantly higher physical demands and strategic challenges during games compared to their younger or semi-professional counterparts [28]. Understanding these fluctuations in external peak requirements could offer several advantages to basketball practitioners: (1) the ability to develop more precise conditioning regimens, (2) the optimization of player rotations during games to maintain optimal physical performance throughout game periods, (3) the development of strategies for prescribing training drills more accurately based on real game reference values, and (4) preparing players to sustain the physical demands of a basketball game when returning from injury. Therefore, the aim of the present study was to investigate the average PDs encountered by professional basketball players across game quarters within three distinct timeframes (30 s, 1 min, and 3 min). Based on prior research conducted with non-elite male players utilizing microtechnology, it was hypothesized that the average [21] and peak values [10,20,22] would likely decrease across the quarters throughout the game.

2. Materials and Methods

2.1. Participants

Professional elite male basketball players classified as Tier 4 (elite/international level) [25,26] (n = 14, mean ± standard deviation: age: 27.8 ± 3.5 y.0; height: 198.1 ± 10.4 cm; body mass: 97.4 ± 11.6 kg) were monitored during 30 games (30 Eurocup/ABA league). Throughout the data collection period, players engaged in an average of 10 h of training per week, comprising 5 basketball sessions and 2 resistance training sessions. Additionally, they played 2 games weekly. Players included in the study were from all positions: guards, forwards, and centers. The team included in the study is based in Slovenia and has played in the ABA League and Eurocup. Eurocup fixtures were scheduled between Tuesday and Friday, while domestic league games were typically held on weekends.

Game samples from each player were only retained in the final analysis if they completed a minimum of 4 min of playing time derived from devices on each quarter. Playing time derived from devices included all stoppages in play, such as free-throws, fouls, and out-of-bounds, but excluded warm-ups, break periods between quarters, time-outs, or time when players were substituted out of the game [9]. Samples obtained from game instances where players accrued less than 4 min of playing time, as determined by data derived from the monitoring devices, were systematically excluded from the final analyses. This exclusion criterion ensured that only substantial periods of active participation were considered, thus minimizing the impact of brief or negligible contributions to the overall PlayerLoad assessment. Additionally, any player who prematurely exited the match due to injury or experienced a cessation of device functionality, whether due to battery depletion or technical malfunctions, was automatically excluded from the dataset corresponding to that specific match. This meticulous approach to data curation aimed to uphold the integrity and reliability of the analytical outcomes by focusing exclusively on instances where players were actively engaged in gameplay for a meaningful duration. In this regard, all players had to compete in at least 50% of the total monitored matches. Overall, 958 game samples across the 14 players were included in the analyses. All subjects gave their informed consent for inclusion before they participated in the study. The study was conducted in accordance with the Declaration of Helsinki, and the protocol was approved by the Ethics Committee of the University of Pais Vasco (UPV/EHU, code M10_2018_027).

2.2. Design

The study was conducted during the 2023–2024 season. Each player wore a monitoring device (S7, Catapult Sports, Melbourne, Australia) inserted into a fitted neoprene vest underneath their regular playing attire. The device was positioned on the upper thoracic spine between the scapulae [29]. Each device contained microsensor technology consisting of an accelerometer (±16 g, 100 Hz), magnetometer (±4.900 μT, 100 Hz), and gyroscope (up to 2000 deg/s, 100 Hz). All players participating in the study were already acquainted with the monitoring technology, having utilized similar devices extensively during both training sessions and competitive games throughout the preceding season. This familiarity ensured a smooth transition into the data collection phase, as players were accustomed to wearing the devices and understanding their functionalities. To maintain consistency and reliability in the recorded data, the devices were activated approximately 20 to 40 min before the commencement of the warm-up phase preceding each game. By initializing the devices ahead of time, any necessary calibration or synchronization processes could be completed without impinging on the players' pre-match routines. Furthermore, to minimize potential discrepancies arising from variations between the individual units of the monitoring devices, players were assigned the same device for the entirety of the study period. In this regard, the same individual took charge of editing all monitored sessions to minimize inter-rater error to its lowest possible extent. This approach effectively mitigated the risk of inter-unit variability in output readings, ensuring that the data obtained remained consistent and comparable across all participants [30].

2.3. Variables

PDs were calculated for PlayerLoad™ (PL) in absolute values and extracted for each player and time window (30 s, 1 min, and 3 min). Research has identified these sample durations as the most practical for consideration in basketball [31]. Average demands were extracted as relative values (Pl·min). PL, a parameter commonly used to measure external load in various sports [6,32–34], is derived from accelerometer data and captures the athlete's accelerations in different planes. However, its definition may vary depending on the manufacturer of the accelerometer device. Recent studies have incorporated PL and have demonstrated its strong correlation with physical and physiological performance [14,35]. Specifically, in the case of basketball, it has the potential to provide a good estimate of the external load of the athlete, as it is a sport that involves a high number of accelerations and decelerations, changes in direction, or explosive efforts. PL was calculated as the square root of the sum of the instantaneous rate of change in acceleration in the three planes of movement (x, y, and z axes) using the following formula [13]: $PlayerLoad^{TM} = \frac{\left[\sqrt{\left(fwd_{t=i+1} - fwd_{t=i}\right)^2} + \sqrt{\left(side_{t=i+1} - side_{t=i}\right)^2} + \sqrt{\left(up_{t=i+1} - up_{t=i}\right)^2}\right]}{100}$, where "fwd" indicates movement in the anteroposterior direction, "side" indicates movement in the medial–lateral direction, "up" indicates vertical movement, and t represents the time.

The average and PDs were determined directly from the Catapult software (OpenField v8, Catapult Innovations, Melbourne, Australia). Peak values were calculated as rolling averages, which is a more precise technique for measuring PD compared to fixed methods [36,37] and has been previously used in basketball research [6,8,31]. After extraction, PDs were input into customized Microsoft Excel (version 16.0, Microsoft Corporation, Redmond, WA, USA) spreadsheets for further analysis.

2.4. Statistical Analysis

The mean and standard deviation (SD) were determined for PL variables across each sample duration (30 s, 1 min, and 3 min). Data distribution normality and sphericity were validated through the Shapiro–Wilk statistic and Levene's Test for homogeneity of variances. A linear mixed model (LMM) with Bonferroni post hoc tests considering significance at $p < 0.05$ was used to compare peak values for each sample duration. Game quarter (4 levels) was entered as a fixed factor, while player (n = 14) was entered as the random term. To identify the magnitude of the differences between quarters, effect sizes (ESs) (Cohen's d) with 95% confidence intervals were calculated. The ES magnitudes of the differences were interpreted as follows: $\leq$0.2, trivial; >0.2, small; >0.6, moderate; >1.2, large; >2.0, very large; and >4.0, nearly perfect [38]. All statistical analyses were conducted using the software jamovi 2.3 (the jamovi project, 2022) for Windows.

3. Results

PL·min across game quarters (Qs) is presented in Figure 1. A decreasing pattern for average values was found among quarters with significant differences ($p < 0.001$) from the first three quarters compared to the last (Q1 vs. Q4, moderate; Q2 vs. Q4, small; Q3 vs. Q4, small).

Table 1 presents the descriptive values of PL for the three sample durations across game quarters and the entire game. Table 2 illustrates the pairwise comparison of PL at different time epochs. For the 30 s sample duration, a significant difference with a *small* effect was observed between game quarters. Regarding the 1 min sample, differences in PL between quarters were significant between Q1 versus Q3 and Q4 ($p < 0.001$, *small* to *moderate*), Q2 with Q3 ($p < 0.001$, *moderate*), and Q3 with Q4 ($p < 0.05$, *small*). For the 3 min sample duration, there was a significant decline in PL from the first three quarters compared to Q4, with a *small* to *moderate* effect. The only non-significant comparison for the 3 min sample duration was between Q2 and Q3.

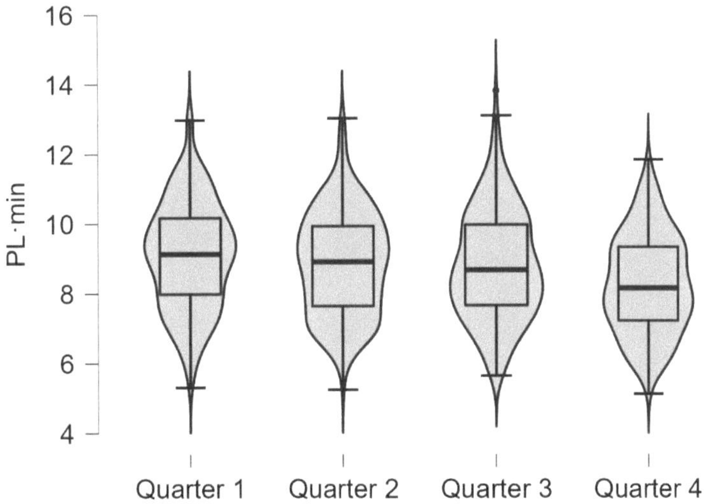

Figure 1. Average values (PL·min) across basketball quarters for elite basketball players.

Table 1. Descriptive values (mean ± standard deviation) in PL between game quarters for each sample duration.

Game Quarters	PL 30 s (AU)	PL 1 min (AU)	PL 3 min (AU)
Q1	10.9 ± 1.6	17.4 ± 3	37.1 ± 6.6
Q2	10.8 ± 1.6	16.9 ± 2.6	35.0 ± 6
Q3	10.4 ± 1.6	16.5 ± 2.8	34.9 ± 6.2
Q4	10.1 ± 1.6	15.8 ± 2.5	33.4 ± 5.5

Note: Q: quarter; PL: PlayerLoad.

Table 2. Pairwise comparison of PL across the three selected timeframes for each quarter of basketball games.

Sample Duration	Effect Size	95% CI	*p*
30 s sample			
Q1 vs. Q2	0.07	(−0.17–0.31)	1.00
Q1 vs. Q3	0.30	(0.05–0.55)	<0.05
Q1 vs. Q4	0.47	(0.22–0.72)	<0.001
Q2 vs. Q3	0.23	(−0.01–0.47)	0.073
Q2 vs. Q4	0.40	(0.16–0.64)	<0.001
Q3 vs. Q4	0.17	(−0.08–0.42)	0.44
1 min sample			
Q1 vs. Q2	0.17	(−0.07–0.41)	0.37
Q1 vs. Q3	0.36	(0.10–0.60)	<0.001
Q1 vs. Q4	0.57	(0.32–0.82)	<0.001
Q2 vs. Q3	0.18	(−0.06–0.42)	0.30
Q2 vs. Q4	0.40	(0.15–0.64)	<0.001
Q3 vs. Q4	0.22	(0.03–0.47)	<0.05

Table 2. *Cont.*

Sample Duration	Effect Size	95% CI	p
3 min sample			
Q1 vs. Q2	0.35	(0.10–0.60)	<0.001
Q1 vs. Q3	0.36	(0.10–0.62)	<0.001
Q1 vs. Q4	0.62	(0.36–0.87)	<0.001
Q2 vs. Q3	0.01	(−0.24–0.26)	1.0
Q2 vs. Q4	0.27	0.02–0.52	<0.05
Q3 vs. Q4	0.26	(0.01–0.51)	<0.05

Note: Q: quarter; CI: confident interval; *p*: *p* value.

4. Discussion

The aim of the present study was to investigate the average physical demands and PDs encountered by professional basketball players across game quarters within three distinct timeframes (30 s, 1 min, and 3 min). This study provides impactful findings for basketball coaches and performance staff, demonstrating a clear decrease in both average and peak external load values for PL across game quarters. Significant differences were observed between the first three quarters and the last quarter, indicating a decline in player activity as the game progressed.

When contrasting our findings with the existing basketball literature, we found similarities with García et al. (2020), who examined male basketball players competing in the Spanish Second Division. They observed a decrease in average physical demands from the first to the fourth quarters, specifically in total distance covered ($p < 0.001$; ES = −1.31) and PL ($p < 0.001$; ES = −1.27) [21]. Regarding peak values, the current research further investigated external PDs across quarters in basketball, focusing on elite junior male players [20], semi-professional male players (Fox et al., 2021), and male basketball players competing in the Spanish Second Division [22]. Their studies also identified significant decreases throughout the game, with the most notable declines in external peak values occurring between the first and fourth quarters for total distance [20,22], PL [10,20,22] and high-speed running [20,22]. However, the differences between quarters in jogging, running, acceleration, and deceleration for all sample durations (30 s and 45 s and 1, 2, and 5 min) were non-significant [20]. These results highlight the general trend of declining average and peak physical demands as the game progresses, indicating that this decrease is independent of the level of competition.

The hypothesis supporting this phenomenon (decline in average and peak values as the game progresses) may be attributed to fatigue-related mechanisms associated with accumulated playing time throughout entire games and prior to intense periods [9]. It may also depend on situational variables such as the team lineup, which can vary due to differences in player capacities, team cohesion, and tactical approaches, as well as the stage of the game (e.g., the game pace may decline during latter periods) [9,39]. Longer sample durations were more sensitive for detecting differences in PDs between quarters, indicating their usefulness in planning game-like conditioning. In contrast, shorter sample durations may reflect situational load demands during live play. This suggests that longer-duration samples can provide insights into conditioning, while shorter-duration intervals can help understand the demands of live play. Overall, these findings underscore the importance of considering both the duration of sampling intervals and the specific quarters of play when analyzing PL dynamics.

4.1. Practical Applications

Our findings can offer a useful practical application for basketball practitioners in several ways. First, training programs can be designed to enhance players' endurance and recovery capacity to sustain performance throughout the game. For instance, training regimens can incorporate drills with elevated PL towards the end of the session, thereby aiming to bolster conditioning levels during periods away from competitive fixtures. Conversely, an alternative approach could involve introducing tasks with diminished PL towards the session's conclusion, more accurately simulating the conditions typically encountered during match play. This approach focuses players on cognitive or tactical aspects, which are often crucial late in the game. The goal is for players to work on strategic aspects in situations of fatigue.

Another application could be strategically rotating players during the game or practice to distribute playing time more equitably and maintain optimal performance. This approach would likely involve pre-match or pre-training planning by coaching staff regarding player participation and rest times. By scheduling strategic rest intervals, coaches can mitigate fatigue and optimize player performance at critical moments.

Finally, it is advisable to use real-time performance tracking technologies to monitor players' physical condition and performance during matches and training sessions. This enables adjustments to participation times based on their physical condition. Overall, understanding how physical demands vary throughout the game can help optimize training, player rotation, and game strategies to maximize team performance across all quarters.

4.2. Limitations

The research conducted presents important strengths, such as the inclusion of elite and professional players, which highlights the high level of the sample and the large sample of matches in international-level competition. Nevertheless, some limitations should be considered when interpreting the current findings. First, variables such as score-line, playing position, fixture congestion, players' role, quality of opposition, tactical aspects, and other factors that could have directly or indirectly influenced the results were not controlled for. Therefore, to advance our understanding of average and peak fluctuations, future research should investigate these factors.

Second, another limitation is the focus on a single variable (PL) due to the challenges associated with installing a local positioning system in elite stadiums, which would allow for measuring positional variables such as distance. Future research conducted with elite or professional players should consider this aspect.

Additionally, a significant limitation is the lack of studies on unexamined populations, such as referees or female players. Research analyzing physical fluctuations has predominantly focused on elite male basketball players [10,20,22]. Incorporating other populations, such as referees who also play a crucial role in the game or female athletes whose playing style and physical demands may differ, could provide a more comprehensive perspective on average and peak fluctuations in sports performance. Therefore, future research should include these populations for a more holistic understanding of the topic.

5. Conclusions

The results of this study show a consistent decrease in both the mean and peak values of external physical demands throughout the quarters of a game in professional basketball players, with the most notable reductions occurring between the initial three quarters and the last quarter. These results suggest that professional basketball players tend to experience fatigue or a reduction in physical demands as the game unfolds, underscoring the importance of managing the players' workload and implementing appropriate strategies to optimize performance throughout the game. Another possible explanation could be that tactical aspects related to game management (such as rotations, playing styles, and defense strategies) during the last quarter influence these physical demands.

Author Contributions: Conceptualization, E.A.-P.-C. and H.S.; methodology, E.A.-P.-C. and H.S.; software, F.U.; formal analysis, H.S.; investigation, J.P.; resources, F.U. and J.P.; data curation, F.U. and J.P.; writing—original draft preparation, E.A.-P.-C. and H.S.; writing—review and editing, A.L.; visualization, H.S.; supervision, E.A.-P.-C.; project administration, F.U. and H.S. All authors have read and agreed to the published version of the manuscript.

Funding: This research was partially funded by the Spanish National Agency of Investigation through the project "Scientific and Technological Support to analyze the Training Workload of Basketball teams according to sex, level of the players and season period" (PID2019-106614GBI00) MCIN/AEI/10.13039/501100011033).

Institutional Review Board Statement: The study was conducted in accordance with the Declaration of Helsinki, and the protocol was approved by the Ethics Committee of the University of Pais Vasco (UPV/EHU, code M10_2018_027).

Informed Consent Statement: Informed consent was obtained from all subjects involved in the study.

Data Availability Statement: The data are not available due to privacy or ethical restrictions.

Conflicts of Interest: The authors declare no conflicts of interest.

References

1. Torres-Ronda, L.; Ric, A.; Llabres-Torres, I.; De las Heras, B.; Schelling, X. Position-Dependent Cardiovascular Response and Time-Motion Analysis during Training Drills and Friendly Matches in Elite Male Basketball Players. *J. Strength Cond. Res.* **2016**, *30*, 60–70. [CrossRef]
2. Impellizzeri, F.M.; Marcora, S.M.; Coutts, A.J. Internal and External Training Load: 15 Years On. *Int. J. Sports Physiol. Perform.* **2019**, *14*, 270–273. [CrossRef]
3. Svilar, L.; Castellano, J.; Jukic, I. Comparison of 5vs5 Training Games and Match-Play Using Microsensor Technology in Elite Basketball. *J. Strength Cond. Res.* **2019**, *33*, 1897–1903. [CrossRef]
4. Sosa, C.; Lorenzo, A.; Trapero, J.; Ribas, C.; Alonso, E.; Jimenez, S.L. Specific Absolute Velocity Thresholds during Male Basketball Games Using Local Positional System; Differences between Age Categories. *Appl. Sci.* **2021**, *11*, 4390. [CrossRef]
5. Berkelmans, D.M.; Dalbo, V.J.; Kean, C.O.; Milanović, Z.; Stojanović, E.; Stojiljković, N.; Scanlan, A.T. Heart Rate Monitoring in Basketball: Applications, Player Responses, and Practical Recommendations. *J. Strength Cond. Res.* **2018**, *32*, 2383–2399. [CrossRef]
6. Alonso, E.; Miranda, N.; Zhang, S.; Sosa, C.; Trapero, J.; Lorenzo, J.; Lorenzo, A. Peak Match Demands in Young Basketball Players: Approach and Applications. *Int. J. Environ. Res. Public Health* **2020**, *17*, 2256. [CrossRef]
7. Bourdon, P.C.; Cardinale, M.; Murray, A.; Gastin, P.; Kellmann, M.; Varley, M.C.; Gabbett, T.J.; Coutts, A.J.; Burgess, D.J.; Gregson, W.; et al. Monitoring Athlete Training Loads: Consensus Statement. *Int. J. Sports Physiol. Perform.* **2017**, *12*, S2-161–S2-170. [CrossRef]
8. Vázquez-Guerrero, J.; Garcia, F. Is it enough to use the traditional approach based on average values for basketball physical performance analysis? *Eur. J. Sport Sci.* **2020**, *21*, 1551–1558. [CrossRef]
9. Pérez-Chao, E.A.; Lorenzo, A.; Scanlan, A.; Lisboa, P.; Sosa, C.; Gómez, M. Higher Playing Times Accumulated Across Entire Games and Prior to Intense Passages Reduce the Peak Demands Reached by Elite, Junior, Male Basketball Players. *Am. J. Men. Health* **2021**, *15*. [CrossRef]
10. Fox, J.L.; Salazar, H.; Garcia, F.; Scanlan, A.T. Peak External Intensity Decreases across Quarters during Basketball Games. *Montenegrin J. Sports Sci. Med.* **2021**, *10*, 25–29. [CrossRef]
11. Fox, J.L.; Conte, D.; Stanton, R.; McLean, B.; Scanlan, A.T. The Application of Accelerometer-Derived Moving Averages to Quantify Peak Demands in Basketball: A Comparison of Sample Duration, Playing Role, and Session Type. *J. Strength Cond. Res.* **2021**, *35*, S58–S63. [CrossRef]
12. Vázquez-Guerrero, J.; Ayala, F.; Garcia, F.; Sampaio, J. The Most Demanding Scenarios of Play in Basketball Competition from Elite Under-18 Teams. *Front. Psychol.* **2020**, *11*, 552. [CrossRef]
13. Pérez-Chao, E.A.; Gómez, M.; Scanlan, A.; Ribas, C.; Trapero, J.; Lorenzo, A. Influence of game and quarter results on external peak demands during games in under-18 years, male basketball players. *Biol. Sport* **2023**, *40*, 365–375. [CrossRef]
14. Pino-Ortega, J.; Rojas-Valverde, D.; Gómez-Carmona, C.D.; Bastida-Castillo, A.; Hernández-Belmonte, A.; García-Rubio, J.; Nakamura, F.Y.; Ibáñez, S.J. Impact of Contextual Factors on External Load During a Congested-Fixture Tournament in Elite U'18 Basketball Players. *Front. Psychol.* **2019**, *10*, 1100. [CrossRef]
15. Pérez-Chao, E.A.; Trapero, J.; Ribas, C.; Sosa, C.; Gómez, M.; Lorenzo, A. External peak demands are not affected by playing venue (home vs away) during official male basketball games. *Int. J. Perform. Anal. Sport* **2023**, *23*, 1–9. [CrossRef]
16. Pérez-Chao, E.A.; Lorenzo, A.; Ribas, C.; Portes, R.; Leicht, A.S.; Gómez, M.A. Influence of analysis focus and playing time on internal average and peak physical demands of professional male basketball players during competition. *RICYDE. Rev. Int. Cienc. Deport.* **2022**, *18*, 155–163. [CrossRef]
17. Vázquez-Guerrero, J.; Vizuete, J.J.; García, F.; Hughes, J.; Croix, M.B.D.S.; Ayala, F. The most demanding scenarios of 5-on-5 modified scrimmage situations in elite basketball. *J. Sports Med. Phys. Fit.* **2021**, *61*, 907–915. [CrossRef]
18. Pérez-Chao, E.A.; Ribas, C.; Gómez, M.A.; Portes, R.; Jiménez, S.L.; Lorenzo, A. Should we train as we compete? Games might be the best scenario to reach the internal peak demands in professional basketball players. *Med. Sport* **2022**, *75*, 45–58. [CrossRef]
19. García, F.; Castellano, J.; Reche, X.; Vázquez-Guerrero, J. Average Game Physical Demands and the Most Demanding Scenarios of Basketball Competition in Various Age Groups. *J. Hum. Kinet.* **2021**, *79*, 165–174. [CrossRef]

20. Pérez-Chao, E.A.; Gómez, M.; Lisboa, P.; Trapero, J.; Jiménez, S.L.; Lorenzo, A. Fluctuations in External Peak Demands across Quarters during Basketball Games. *Front. Physiol.* **2022**, *13*, 868009. [CrossRef]
21. García, F.; Vázquez-Guerrero, J.; Castellano, J.; Casals, M.; Schelling, X. Physical Demands between Game Quarters and Playing Positions on Professional Basketball Players during Official Competition. *J. Sports Sci. Med.* **2020**, *19*, 256–263. [PubMed]
22. García, F.; Salazar, H.; Fox, J.L. Differences in the Most Demanding Scenarios of Basketball Match-Play between Game Quarters and Playing Positions in Professional Players. *Montenegrin J. Sports Sci. Med.* **2022**, *11*, 15–28. [CrossRef]
23. García, F.; Schelling, X.; Castellano, J.; Martín-García, A.; Pla, F.; Vázquez-Guerrero, J. Comparison of the most demanding scenarios during different in-season training sessions and official matches in professional basketball players. *Biol. Sport* **2022**, *39*, 237–244. [CrossRef] [PubMed]
24. Petway, A.J.; Freitas, T.T.; Calleja-González, J.; Leal, D.M.; Alcaraz, P.E. Training load and match-play demands in basketball based on competition level: A systematic review. *PLoS ONE* **2020**, *15*, e0229212. [CrossRef]
25. McKay, A.K.; Stellingwerff, T.; Smith, E.S.; Martin, D.T.; Mujika, I.; Goosey-Tolfrey, V.L.; Sheppard, J.; Burke, L.M. Defining Training and Performance Caliber: A Participant Classification Framework. *Int. J. Sports Physiol. Perform.* **2022**, *17*, 317–331. [CrossRef] [PubMed]
26. Swann, C.; Moran, A.; Piggott, D. Defining elite athletes: Issues in the study of expert performance in sport psychology. *Psychol. Sport Exerc.* **2015**, *16*, 3–14. [CrossRef]
27. Scanlan, A.T.; Tucker, P.S.; Dascombe, B.J.; Berkelmans, D.M.; Hiskens, M.I.; Dalbo, V.J. Fluctuations in Activity Demands Across Game Quarters in Professional and Semiprofessional Male Basketball. *J. Strength Cond. Res.* **2015**, *29*, 3006–3015. [CrossRef]
28. Ben Abdelkrim, N.; Castagna, C.; Jabri, I.; Battikh, T.; El Fazaa, S.; El Ati, J. Activity Profile and Physiological Requirements of Junior Elite Basketball Players in Relation to Aerobic-Anaerobic Fitness. *J. Strength Cond. Res.* **2010**, *24*, 2330–2342. [CrossRef]
29. Hodder, R.W.; Ball, K.A.; Serpiello, F.R. Criterion Validity of Catapult ClearSky T6 Local Positioning System for Measuring Inter-Unit Distance. *Sensors* **2020**, *20*, 3693. [CrossRef]
30. Castellano, J.; Casamichana, D.; Calleja-González, J.; Román, J.S.; Ostojic, S.M. Reliability and Accuracy of 10 Hz GPS Devices for Short-Distance Exercise. *J. Sports Sci. Med.* **2011**, *10*, 233–234.
31. Pérez-Chao, E.A.; Portes, R.; Gómez, M.; Parmar, N.; Lorenzo, A.; Sáiz, S.L.J. A Narrative Review of the Most Demanding Scenarios in Basketball: Current Trends and Future Directions. *J. Hum. Kinet.* **2023**, *89*, 231–245. [CrossRef] [PubMed]
32. Aughey, R.J. Applications of GPS Technologies to Field Sports the Validity of GPS for the Measurement. *Int. J. Sports Physiol. Perform.* **2011**, *6*, 295–310. [CrossRef] [PubMed]
33. Holme, B. Wearable Microsensor Technology to Measure Physical Activity Demands in Handball. Master's Thesis, Norwegian School of Sport Sciences, Oslo, Norway, 2015.
34. Schelling, X.; Torres, L. Accelerometer Load Profiles for Basketball-Specific Drills in Elite Players. *J. Sports Sci. Med.* **2016**, *15*, 585–591. [PubMed]
35. Vázquez-Guerrero, J.; Jones, B.; Fernández-Valdés, B.; Moras, G.; Reche, X.; Sampaio, J. Physical Demands of Elite Basketball during an Official U18 International Tournament. *J. Sports Sci. Med.* **2019**, *37*, 2530–2537. [CrossRef] [PubMed]
36. Oliva-Lozano, J.M.; Martín-Fuentes, I.; Fortes, V.; Muyor, J.M. Differences in worst-case scenarios calculated by fixed length and rolling average methods in professional soccer match-play. *Biol. Sport* **2021**, *38*, 325–331. [CrossRef]
37. Cunningham, D.J.; Shearer, D.A.; Carter, N.; Drawer, S.; Pollard, B.; Bennett, M.; Eager, R.; Cook, C.J.; Farrell, J.; Russell, M.; et al. Assessing worst case scenarios in movement demands derived from global positioning systems during international rugby union matches: Rolling averages versus fixed length epochs. *PLoS ONE* **2018**, *13*, e0195197. [CrossRef]
38. Hopkins, W.; Marshall, S.; Batterham, A.; Hanin, J. Progressive statistics for studies in sports medicine and exercise science. *Med. Sci. Sports Exerc.* **2009**, *41*, 3. [CrossRef]
39. Bradley, P.S.; Noakes, T.D. Match running performance fluctuations in elite soccer: Indicative of fatigue, pacing or situational influences? *J. Sports Sci.* **2013**, *31*, 1627–1638. [CrossRef]

sensors

Article

Mechanical Determinants of Sprinting and Change of Direction in Elite Female Field Hockey Players

Alejandro Bustamante-Garrido [1,2,3,4], Mikel Izquierdo [2], Bianca Miarka [5], Ariel Cuartero-Navarrete [6], Jorge Pérez-Contreras [1,3,7], Esteban Aedo-Muñoz [3,4,5,8,*] and Hugo Cerda-Kohler [3,5,9,*]

1 Escuela de Ciencias del Deporte y la Actividad Física, Facultad de Salud, Universidad Santo Tomás, Santiago 8370003, Chile; abustamante7@santotomas.cl (A.B.-G.); joperezc@gmail.com (J.P.-C.)
2 Navarrabiomed, Hospitalario Universitario de Navarra (HUN), Universidad Pública de Navarra (UPNA), IdiSNA, 31008 Pamplona, Spain; mikel.izquierdo@unavarra.es or mikel.izquierdo@gmail.com
3 Departamento de Educación Física, Deportes y Recreación, Facultad de Artes y Educación Física, Universidad Metropolitana de Ciencias de la Educación, Santiago 7760197, Chile
4 Laboratorio de Biomecánica Deportiva, Unidad de Ciencias Aplicadas al Deporte, Instituto Nacional de Deportes, Ministerio del Deporte de Chile, Santiago 7750000, Chile
5 Laboratory of Psychophysiology and Performance in Sports and Combats, Postgraduate Program in Physical Education, School of Physical Education and Sport, Federal University of Rio de Janeiro, Rio de Janeiro 21941-599, Brazil; miarkasport@hotmail.com
6 Federación Chilena de Hockey Sobre Césped, Santiago 7750000, Chile; arielcuartero87@gmail.com
7 Escuela de Doctorado, Universidad de las Palmas de Gran Canaria (EDULPGC), 35001 Las Palmas, Spain
8 Escuela de Ciencias de la Actividad Física, el Deporte y la Salud, Facultad de Ciencias Médicas, Universidad de Santiago de Chile, Santiago 9170022, Chile
9 Unidad de Fisiología del Ejercicio, Centro de Innovación, Clínica MEDS, Santiago 7691236, Chile
* Correspondence: esteban.aedo@usach.cl (E.A.-M.); hugorck@gmail.com (H.C.-K.)

Citation: Bustamante-Garrido, A.; Izquierdo, M.; Miarka, B.; Cuartero-Navarrete, A.; Pérez-Contreras, J.; Aedo-Muñoz, E.; Cerda-Kohler, H. Mechanical Determinants of Sprinting and Change of Direction in Elite Female Field Hockey Players. *Sensors* **2023**, *23*, 7663. https://doi.org/10.3390/s23187663

Academic Editors: Bruno Gonçalves, Bruno Emanuel Nogueira Figueira and Diogo Alexandre Martins Coutinho

Received: 6 August 2023
Revised: 29 August 2023
Accepted: 31 August 2023
Published: 5 September 2023

Abstract: Profile determination in field hockey is critical to determining athletes' physical strengths and weaknesses, and is key in planning, programming, and monitoring training. This study pursued two primary objectives: (i) to provide descriptive data on sprinting, deceleration, and change of direction (COD) abilities and (ii) to elucidate the mechanical variables that influence sprint and COD performance in elite female field hockey players. Using radar and time-gate technology, we assessed performance and mechanical data from 30 m sprinting, deceleration, and COD tests for 26 elite female hockey players. A machine learning approach identified mechanical variables related to sprint and COD performance. Our findings offer a framework for athlete categorization and the design of performance-enhancing training strategies at the international level. Two pivotal mechanical variables—relative maximum horizontal force (F0) and maximum velocity (Vmax)—predominantly influence the times across all tested distances. However, the force–velocity profile (FVP) and horizontal deceleration do not influence the variance in the COD test outcomes. These insights can guide the design, adjustment, and monitoring of training programs, assisting coaches in decision making to optimize performance and mitigate injury risks for female hockey players.

Keywords: physical and physiological analysis; training and game monitoring; motion analysis

1. Introduction

Decision making is essential in any sport [1], particularly in dynamic team sports such as field hockey, volleyball, soccer, rugby, and basketball. In this process, coaches and staff play a key role in planning, programming, and monitoring training [2]. In order to succeed, it is crucial to know the sport's profile, including biomechanical, physiological, and technical-tactical factors. Therefore, profile determination is critical to determining athletes' physical strengths and weaknesses and fundamental to making decisions about the training process [3].

Field hockey is an invasive team sport featuring many offensive and defensive abilities mixed with intermittent high-intensity running [2,4], which represent significant percentages of the entire game (12 to 26%) [5]. This profile is characterized by movements such as high-velocity running, jumping, changes of direction, kicks, and hits, demonstrating the power generated by the athlete [6]. Having high-intensity running capacity and maintaining these actions consistently in a game or season is an essential aspect of team sports [4] and can mean victory [5]. Similarly, changing direction (COD) is crucial, enabling competitors to evade opponents and gain advantageous positions [7,8]. COD is an implicit skill in agility, including acceleration, deceleration, and decision-making situations [9]. Accordingly, several aspects of sprinting and COD must be considered in field hockey athletes' training process, such as duration, intensity, time spent at different intensities, and physical and mechanical determinants.

Regarding sprinting, the average duration in elite matches is 1.8 ± 0.4 s (s), and the longest sprints last 4.1 ± 2.1 s [10]. Additionally, sprints typically cover distances ranging from 10 to 20 meters (m) [11] and account for 1.4–2.1% (73–112 s) of the total game time [12]. Sprints are also performed at between 18 and 24 km/h, representing 91 ± 4% of the players' maximum speed [12,13]. The applied horizontal force is one of the most relevant variables when executing a sprint [14–16]. The force–velocity profile (FVP) method allows the identification of mechanical variables such as the maximum horizontal force (F0), maximum velocity (V0), and maximum horizontal power (Pmax) applied during the sprint [17]. This method is a valuable tool for coaches and physical trainers in the determination of the physical profile of the athlete. Several studies have studied the profile of different sports, such as soccer [15,18], athletics [19], ice hockey [20,21], basketball, and tennis [22]. However, evidence of the force–velocity profile and the mechanical determinants of the sprint in elite female field hockey is lacking.

The ability to perform quick sprints and efficiently change direction is crucial for sport performance. On average, male players perform 1148 ± 128.9 change of direction movements during elite field hockey matches [23]. Furthermore, another study revealed 512 ± 69 acceleration–deceleration actions in elite female players [24]. Strength, power, speed, and technical ability determine the athlete's ability to change direction and their acceleration–deceleration actions [9]. In addition, fast decelerations are crucial for quick change of direction (COD) movements performed during team sports [25,26]. However, the associations between mechanical factors, such as FVP and deacceleration, and COD in elite female field hockey has not been investigated.

The identification of the mechanical variables of sprinting (at different distances) and COD allows the identification of the athlete's profile to optimize performance and determine risk factors for injury [27] and post-injury monitoring [28]. This study has two objectives: to present sprint, change of direction, and deceleration reference values, and to determine the influences of mechanical variables of the FVP and deceleration on sprint acceleration performance (at 5, 15, and 30 m) and the ability to change direction in elite female field hockey athletes. We hypothesized that sprint mechanical variables and deceleration significantly explain sprint and COD performance in elite female field hockey players.

2. Materials and Methods

2.1. Participants

Twenty-eight female players in the Chilean national team participated in this study. Table 1 shows participants' characteristics (field hockey training experience, 18.9 ± 4.7 years; time representing the Chilean national team, 6.2 ± 4.9 years). Athletes voluntarily agreed to participate in this study, signing an informed consent form. The assessments were carried out following the Declaration of Helsinki standards. The tests were conducted in a competitive period, one week before a series of official world-ranking matches.

Table 1. Description of the subjects (mean ± standard deviation).

	Body Mass (kg)	Height (cm)	Age (Years Old)
Defenders (8)	64.8 ± 6.8	166.1 ± 2.1	27.0 ± 3.7
Midfields (11)	62.3 ± 5.0	166.2 ±4.5	24.8 ± 4.6
Forwards (7)	61.4 ± 2.7	166.4 ± 4.3	24.7 ± 4.3
Total (26)	63.1 ± 5.3	166.2 ± 4.1	25.38 ± 4.2

2.2. Design

The testing was conducted between 07:30 a.m. and 09:00 a.m. in Chile's official field hockey arena. Before starting, the protocol of each evaluation was detailed to the participants after each athlete performed a general warm-up similar to the one they performed during their physical preparation sessions. The warm-up lasted 15 min, including low-moderate intensity cardiovascular exercises, mobility, and short-distance sprints. After the standardized warm-up, subjects performed two maximal sprints on an indoor track. They all performed a linear run with a maximum acceleration of 30 m to determine the force–velocity profile (FVP). This profile represents the force–velocity and power–velocity relationship that the neuromuscular system of the lower extremities is capable of generating [29]. Only the best times were considered for data analysis for the two trials.

2.2.1. Speed-Acceleration Evaluation 30 m

We evaluated the maximum acceleration in 30 m on a field hockey field. For this, the Stalker ATS II model radar (Applied Concepts, Dallas, TX, USA; accuracy + 1.61 km/h, sampling 46.9 Hz) was located on a tripod 10 m from the starting line and at the height of 1 m to align with the location of the center of mass (CM) of the subject [30]. Radar technology relies on using sonic waves to ascertain an object's distance. Additionally, velocity can be precisely measured through the Doppler effect (variation in wave frequency as an object approaches or recedes) [31]. The radar was operated from a computer to avoid the variability produced by manual operation [32]. Athletes were instructed to initiate with no backward step, performing two 30 m maximal sprints from a standing staggered-stance start, with at least 5 min passive recovery between sprints. Sprint performance (split times 0–5, 0–15, and 0–30 m) and mechanical outputs were computed for the best time trial.

2.2.2. Deceleration

Athletes used the same start protocol and radar technology for the horizontal sprint test and sprinted maximally over 30 m before performing a maximal horizontal deceleration. As previously described [25], we defined the beginning of the deceleration phase as the time point immediately following the maximum velocity achieved during the 30 m sprint. The end of the deceleration phase was established as the lowest velocity following maximum velocity. We used early deceleration phases for analysis using the time point associated with 50% maximum velocity.

2.2.3. Agility Test

Athletes performed the change of direction ability (CODA) test to assess agility. The test consists of performing a forward sprint of 10 m and two lateral runs of 8 m, with the final line of 10 m as a reference. Finally, the athlete returns to the starting line to complete the test (10-8-8-8-10 = 36 m) [33]. A Witty time gate (Microgate, Bolzano, Italy) placed on a tripod at a height of 1 m was used to determine the time. To prevent athletes from activating gates with their arms, they were positioned 50 cm behind the light gate. Each athlete performed two attempts, separated by 4 min of rest. The best record was used for the analysis.

2.2.4. Horizontal Force Velocity Profile

The speed–time data obtained by radar were loaded into the Excel® spreadsheet created by Morin and Samozino [34]. The spreadsheet calculates the maximum horizontal force used during the sprint (F0), maximum velocity (V0), and the maximum horizontal power output (Pmax). In addition, we obtained the proportion of the total force produced by the lower limbs on the floor that is applied horizontally (RFmax) and the rate of decrease in horizontal force as speed increases (D_{RF}). This method uses the fundamental laws of movement to obtain the force–velocity relationship using the athlete's speed and body mass [32]. The use of radar for these purposes was validated using force platforms (absolute bias 3–7%) [17]. Regarding reliability, the mean typical error is small (CV $\leq$ 8.4%) for all kinetic and kinematic variables [35], making it a valuable field tool for determining these variables. Briefly, the net horizontal anteroposterior GRF (F_H) applied to the body center of mass (CM) can be modeled over time as follows [17]:

$$F_H(t) = m \cdot a_H(t) + F_{aero}(t)$$

where m is the runner's body mass (in kg) and $F_{aero}(t)$ is the aerodynamic drag that must be overcome in a sprint, which is proportional to the square of the velocity of air relative to the runner:

$$F_{aero}(t) = k \cdot (v_H(t) - v_w)^2$$

where v_w is the wind velocity (if any) and k is the runner's aerodynamic friction coefficient.

Regarding vertical direction, during the acceleration phase, the runner's body CM goes up from the starting position to the upright running position and does not change from one complete step to another. Therefore, using the fundamental laws of dynamics in the vertical direction, the mean net vertical ground reaction forces (F_V) applied to the body CM over each complete step can be modeled over time as being equal to body weight:

$$F_V(t) = m \cdot g$$

where g is the gravitational acceleration (9.81 m/s^2).

The mechanical effectiveness of force application during running could be quantified over each support phase or step by the ratio (RF in %) of F_H to the corresponding total resultant ground reaction forces (F_{Res}, in N) and the entire acceleration phase by the slope of the linear decrease in RF when velocity increases (D_{RF}, in %/s/m):

$$RF = \frac{F_H}{F_{Res}} \cdot 100 = \frac{F_H}{\sqrt{{F_H}^2 + {F_V}^2}} \cdot 100$$

Because the starting block phase (push-off and following aerial time) lasts between 0.5 and 0.6 s [28,29], occurring for an average time of ~0.3 s, RF and DRF can be computed from F_H and F_V values modeled for t > 0.3 s.

2.2.5. Statistical Analysis

The descriptive data are presented as mean and standard deviation (mean ± SD). Minimum and maximum (median, quartile 25 (Q_{25}) and 75 (Q_{75})) values for each variable are also reported for better interpretation of the data. The study utilized a machine learning approach to examine the relationships between our target variable "x" and a set of predictor variables "y". Linear regression is a statistical technique that predicts the outcome of a response variable using several explanatory variables and is used to model the linear relationship between explanatory variables and response variables. The model assumes the absence of multicollinearity, which means that the explanatory variables are not highly correlated. This study aimed to determine which independent variables most significantly explain the variation in our target variable. Twelve variables were used to build the models explaining the result of sprint acceleration and agility.

The study's target (i.e., dependent) variable was time at a different distance in the FVP and the CODA test. All other columns in the dataset were treated as independent variables. In order to avoid multicollinearity, which can bias the interpretation of regression coefficients, a correlation matrix of the independent variables was calculated. In general, if the absolute value of the Pearson correlation coefficient is >0.8, collinearity is likely to exist [36,37]. Thus, variables with a correlation coefficient higher than 0.80 with another variable were removed.

The analysis employed fivefold cross-validation [38,39] to mitigate overfitting and bias, especially given the challenges posed by small datasets. Briefly, fivefold cross-validation divides the dataset into five parts, or "folds". For each iteration, the model is trained using four folds and validated on the remaining fold. This process is repeated five times, ensuring each fold is the validation set once. To guarantee that all features equally influenced the models, we standardized them to have a mean of 0 and a standard deviation of 1 through standard scaling. Such a step is crucial as variables with different scales can skew the model fit, potentially leading to biased coefficient estimations.

The study adopted a comprehensive approach to model fitting [40] using three linear regression models: Ordinary Least Squares (OLS), Ridge Regression, and Lasso Regression. The OLS regression attempted to minimize the sum of the squared residuals. This model provided an initial understanding of the relationship between the predictor and target variables without penalty imposed on the coefficients. Ridge and Lasso's regressions introduced a level of bias into the coefficient estimates to manage multicollinearity and improve model interpretability. Ridge Regression includes an L2 penalty that shrinks the coefficients of correlated predictors. Lasso Regression utilizes an L1 penalty that can shrink some coefficients to zero, thus performing feature selection.

A parameter alpha controls the degree of bias or regularization in Ridge and Lasso regressions. Higher alpha values increase the penalty term and thus shrink the coefficients towards zero, effectively simplifying the model. Conversely, an alpha of zero resembles the OLS regression. Therefore, choosing the correct alpha value is critical to balance the model's complexity and predictive power. The study used a range of potential alpha values to choose the optimal alpha: [0.0001, 0.001, 0.01, 0.1, 1, 10]. The analysis employed RidgeCV and LassoCV, which use cross-validation to select the best alpha that gives the best predictive performance on unseen data [41].

Once the models were fitted with the training data, their performance was determined using the coefficient of determination (R^2), describing the proportion of the dependent variable's variance explained by the independent variables. The best explanatory model selection was primarily based on the R^2 from the test dataset, which reflects the model's performance on unseen data. Following the selection of the best explanatory model, the model's performance and stability of the coefficients were evaluated. This involved the calculation of Mean Absolute Error (MAE), Mean Squared Error (MSE), and Root Mean Squared Error (RMSE). These metrics provided a quantitative measure of how our model can explain the actual values [41].

The predicted versus actual values for the training and test datasets were plotted to facilitate interpretation. An identity line was also included to represent perfect predictions, providing a clear visual guide to interpret the model's performance. To ensure the robustness of our linear regression model, we rigorously examined its underlying assumptions. The normality of residuals was assessed using the Shapiro–Wilk Test, complemented by a QQ plot for visual verification of data conformity to a normal distribution. Furthermore, we verified the homogeneity of variance across levels of the independent variables, employing a visual check via a scatter plot of predicted values against residuals and a statistical assessment using the Breusch–Pagan test. Statistical significance was set at $p < 0.05$. Data analysis was performed in Python 3.9 programming language using the packages "pandas", "numpy", "matplotlib", "sklearn", "statsmodels", and "scipy.stats".

3. Results

3.1. Descriptive Values

The descriptive analysis of variables associated with the sport contributes to updating the knowledge used to monitor the training process [1]. In addition, the data could be used to compare athletes' results with international values or similar categories (benchmarking).

3.1.1. Sprint Variables

To determine split times using radar technology for 0–5 m, 0–15 m, and 0–30 m, athletes performed two 30 m maximal sprints. Table 2 shows the times at each distance for the best trial. The times were 1.60 s ± 0.02 s for 0–5 m, 3.21 s ± 0.14 s for 0–15 m, and 5.30 s ± 0.21 s for 0–30 m.

Table 2. Descriptive values of times in 5 m, 15 m, and 30 m, mechanical variables of the horizontal profile, times in the agility test, and deceleration (median, quartile 25 (Q_{25}) and 75 (Q_{75})).

	Q_{25}	Median	Q_{75}
5-m (s)	1.55	1.60	1.65
15-m (s)	3.11	3.21	3.26
30-m (s)	5.18	5.27	5.42
F0 (N/kg)	5.34	5.72	6.05
V0 (m/s)	7.5	7.97	8.34
Pmax (W/kg)	10.61	11.51	12.01
FV profile (N/s/m)	−0.79	−0.73	−0.67
RF max (%)	41.28	43.4	44.83
DRF (%/s/m)	−7.33	−6.85	−0.62
Deceleration (m/s^2)	3.00	3.29	3.69
CODA (s)	9.01	9.23	9.50

F0: maximal theoretical horizontal force; V0: maximal theoretical velocity; Pmax: maximal horizontal power; FVslope: the slope of the linear F-V relationship; DRF: decrease in the ratio of horizontal-to-resultant force; RFmax: maximal ratio of horizontal-to-resultant force. CODA: change of direction ability.

3.1.2. Deceleration and Change of Direction

To describe early deceleration using radar technology, athletes sprint maximally over 30 m, followed by a maximal horizontal deceleration (Table 2). Average values for early deceleration were −3.37 ± 0.87 m/s^2. The time employed during the CODA test was recorded using time gates to assess agility. The mean time was 9.28 ± 0.30 s.

3.1.3. Mechanical Variables

The speed–time data obtained by radar technology were loaded into an Excel® spreadsheet to determine the mechanical outputs from the best trial of the two 30 m maximal sprints. Table 2 shows the quartile values of the variables from the FVP at 30 m. The mean ± SD values were 5.72 ± 0.49 N/kg for relative F0, 7.93 ± 0.47 m/s for V0, 11.35 ± 1.21 W/kg for relative Pmax, −0.72 ± 0.07 N/s/m for FV slope, 43.2 ± 2.0% for RFmax, and −6.75 ± 0.13 %/s/m for D_{RF}.

3.2. Mechanical Variables Determining Sprint and Change of Direction Performance

A machine learning approach with three linear regression models was used to determine the best explanatory model for the time at different distances in the FVP and the CODA test. The best explanatory model selection was primarily based on the R^2 from the test dataset, reflecting the model's performance on unseen data.

3.2.1. Five Meter Sprint Performance

In many team sports, sprinting occurs over short distances and initial acceleration (0–10 m) is critical to performance [42]. The Lasso regression was the best model for explaining the 5 m time by the mechanical variables relative F0 and Vmax (R^2 = 0.72; MAE: 0.025; MSE:

0.001; RMSE: 0.033; intercept: 2.8204; relative F0 coefficient: −0.1493; Vmax coefficient: −0.0481; Figure 1).

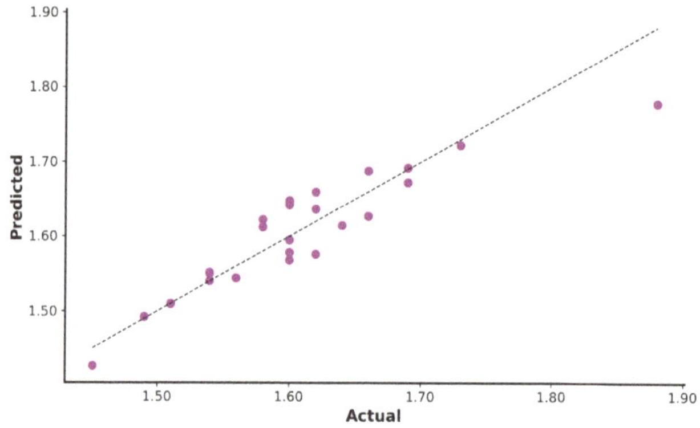

Figure 1. Regression plot of actual versus predicted for 5 m sprint. The black dotted line represents the linear regression function.

3.2.2. Fifteen Meter Sprint Performance

The mechanical variables relative F0 and Vmax determine the time taken to cover a distance of 15 m. The Lasso regression model best explains the 15 m time (R^2 = 0.82; MAE: 0.028; MSE: 0.001; RMSE: 0.038; intercept: 5.6568; relative F0 coefficient: −0.2172; Vmax coefficient: −0.1611; absolute F0 coefficient: 1.5058; Figure 2).

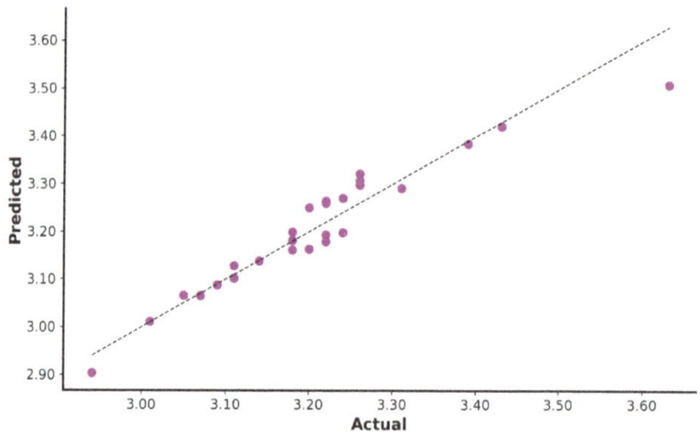

Figure 2. Regression plot of actual versus predicted for 15 m sprint. The black dotted line represents the linear regression function.

3.2.3. Thirty Meter Sprint Performance

The average sprint distance in field hockey players is less than 20 m [43]. However, we consider it relevant to incorporate the 30 m test for occasions where the athlete must cover longer distances and to determine the FVP profile. Similar to the 5 m and 15 m times, the variables relative F0 and Vmax explained the time taken to cover a distance of 30 m. The Lasso regression model was the best explanatory model for the dependent variable (R^2 = 0.93; MAE: 0.027; MSE: 0.001; RMSE: 0.035; intercept: 9.6188; relative F0 coefficient: −0.2514; Vmax coefficient: −0.3827; Figure 3).

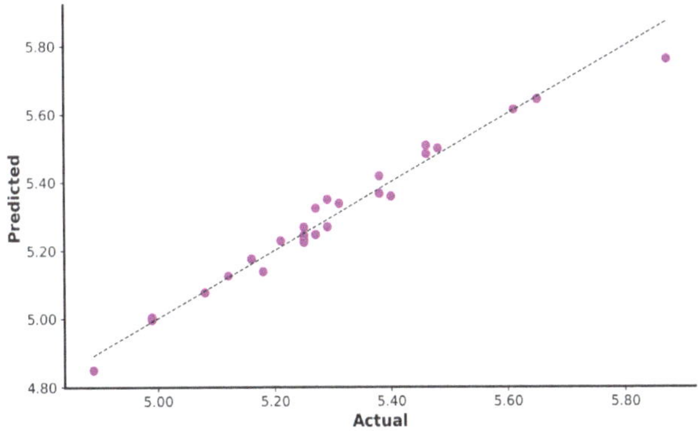

Figure 3. Regression plot of actual versus predicted for 30 m sprint. The black dotted line represents the linear regression function.

3.2.4. Change of Direction

Change of direction performance, assessed through the CODA test, is not significantly explained by the FV profile variables and deceleration ($R^2 = 0.04$).

4. Discussion

The main findings of this study are separated into two components: (i) the descriptive values of a sprint, change of direction, and deacceleration, and (ii) the mechanical variables that determine sprint acceleration performance over short distances and change of direction in elite female field hockey players.

4.1. Descriptive Values

According to our first aim, our results show that the time to run 5 m is longer than that of elite team sport female players (0.99 s vs. 1.60 s) [5]. Regarding 30 m, elite female field hockey players cover the distance in 5 s [44], 0.30 s less than the mean of the athletes presented in our study. In addition, a recent study [45] shows values for the times used to run 5, 15, and 30 m in sports similar to field hockey, including female soccer. The times in the three distances are lower than those in our sample (5 m: 1.52 vs. 1.60, 15 m: 3.09 vs. 3.21, and 30 m: 5.16 vs. 5.30 s). These descriptive values allow for categorization of athletes and defining training strategies to improve performance at the international level.

Little evidence exists about the horizontal FVP in female field hockey. A recent study [46] reported the FV profile values of 31 field hockey players (15 males and 16 females). The average values obtained were F0 = 6.88 N/kg, V0 = 7.69 m/s^2, and Pmax = 13.19 W/kg. Compared with our results, these values are 20.3% higher for F0, 3.0% lower for V0, and 16.2% higher for Pmax. However, direct comparison is challenging due to including male and female participants in their study. Another study describes the values of Norwegian elite athletes from different team sports, such as handball and women's soccer [47]. F0, V0, and Pmax variables are significantly higher than our sample's results (~25.4% higher). Conversely, another study [45] showed that soccer and basketball players present Pmax values of 12.6 and 11.4 (W/kg), similar to those measured in our study. These results highlight the need for further research to establish reference values that allow practitioners to guide training and monitor key performance variables in field hockey.

To the best of the authors' knowledge, this study is the first to show CODA test values in elite field hockey players. Accordingly, there are no specific reference values for female players. This test was initially designed to assess professional football referees' change of direction abilities [48]. FIFA established a reference value of 10 s for referees in the international category. However, our research observed that the average time is lower than the above reference value, with an average time of 9.28 s. Thus, our preliminary results suggest that athletes demonstrate agility performance that exceeds the benchmark value established for international-level soccer referees. Soccer referees perform actions similar to those in field hockey, such as running, sprinting, side-stepping, and running backward [49]. However, the physical space, top speed, and number and magnitude of accelerations and decelerations occurring in field hockey differ from those of soccer referees [12]. Accordingly, further investigations and establishment of specific reference values for women's field hockey are necessary to comprehensively evaluate agility in this sport.

Regarding early deceleration, only a few studies present values in team sports such as soccer, rugby, and netball [25]. Those values are higher than those in our study (-3.83 vs. -3.37 m/s^2). A similar study in team sports showed values of -3.92 m/s^2 [50], suggesting that the female hockey athletes of our study exhibit lower deceleration rates. However, these values represent mixed values for men and women. Therefore, the differences could be due to the gender of the athletes (mainly males), and not to deceleration performance per se. More information on deceleration in field hockey is needed to determine reference values that can be used as a benchmark for coaches and trainers. Finally, training strategies must enhance athletes' ability to dissipate braking loads, improving the muscle functions as a shock absorber (energy attenuation) [51]. This approach will lead to new developments in injury mitigation and physical development strategies in team sports [52].

4.2. Mechanical Variables

The most relevant mechanical variables determining the times for all the distances studied are relative F0 and Vmax, which partially agree with the hypotheses of this study. Our findings align with Hicks et al. [46], who reported that the same mechanical variables explain 94% of the time taken to cover 30 m. Other studies have also investigated mechanical variables impacting sprint performance over various distances, with speed and applied force emerging as crucial factors [53,54]. The logical relationship between applied force and sprint velocity is underpinned by horizontal velocity's dependency on ground reaction forces (GRFs) applied in minimal time [55]. The horizontal component of GRF (propulsive impulse) accounts for 57% of the variation in maximum sprint velocity [56]. Our results corroborate the importance of strength training, mainly horizontally applied force, for enhancing sprint capabilities and the need for shorter contact times. Thus, the specific application of force emerges as a decisive factor in sprint performance, surpassing the magnitude of force applied [57]. This challenges field hockey coaches and practitioners, as players must execute maximum-speed sprints while wielding a stick.

The ability to change direction depends on various mechanical variables such as ground reaction force, impulse, velocity, and braking forces [7,58,59]. However, our results showed that mechanical variables derived from FVP and horizontal deceleration do not significantly explain the variance of performance in the CODA test. Our results do not agree with our hypothesis and align with similar results that found low relationships between mechanical variables and female soccer players' ability to change direction [60]. The FVP profile only identifies horizontally applied force during maximum speed rather than specific braking forces (i.e., horizontal and vertical forces) relevant during COD. Furthermore, the ability to decelerate also depends on variables such as muscle strength (concentric, eccentric, isometric) and technical components [61], which are aspects not addressed in our study.

A limitation of our study is that the frequency sampling of the Stalker radar is 46.9 Hz; however, the literature suggests a frequency between 50 to 250 Hz should be used in motion capture systems [62]. In addition, estimates of variables related to the FVP have been criticized, especially those related to power output. Treating scalar quantities (e.g., power) as a vector could be inappropriate in biomechanics, and vector quantities as impulses could be counted as causative factors in performance [63].

Obtaining the data in a competitive period is a strength of our study, allowing the construction of a valid performance profile. Nevertheless, we recommend assessing the mechanical variables in different general and competitive training periods because training status, training characteristics, responsiveness to training, and nutrition, among other factors, could modify these variables [64]. Finally, control variables such as hormonal profile and anthropometrics could be related to mechanical changes in the training load.

5. Conclusions

Our findings highlight the crucial role of understanding female hockey players' mechanical characteristics of sprinting and COD abilities. By identifying these parameters, researchers and practitioners gain a powerful tool for planning, monitoring, or adjusting training programs. More than just a set of performance metrics, this information provides a comprehensive athlete profile. Such a profile paves the way for individualized training regimes tailored to enhance specific mechanical variables, distances, and skills that need improvement. By refining these individualized training loads, practitioners can support decision making in dynamic team sports. This optimizes performance and plays a crucial role in injury prevention. Consequently, this study is key in setting a benchmark for female hockey training practices, emphasizing a data-driven approach to enhance athletic performance and safety.

Author Contributions: Conceptualization; A.B.-G., H.C.-K. and E.A.-M.; methodology; A.B.-G., H.C.-K. and A.C.-N.; formal analysis; A.B.-G., H.C.-K. and B.M., supervision; M.I. and B.M.; resources; A.B.-G., A.C.-N. and J.P.-C.; writing—review and editing; A.B.-G., H.C.-K. and M.I. All authors have read and agreed to the published version of the manuscript.

Funding: This research received no external funding.

Institutional Review Board Statement: This study does not have an ethical review because the assessments performed are part of the team's test battery in preparation for the World Cup, although all tests were performed in accordance with the Declaration of Helsinki.

Informed Consent Statement: Informed consent was obtained from all subjects involved in the study.

Data Availability Statement: Data are available for research upon request to the corresponding author.

Acknowledgments: The authors acknowledge the staff and athletes of "Federación Chilena de Hockey sobre Césped".

Conflicts of Interest: The authors declare no conflict of interest.

References

1. Houtmeyers, K.C.; Jaspers, A.; Figueiredo, P. Managing the training process in elite sports: From descriptive to prescriptive data analytics. *Int. J. Sports Physiol. Perform.* **2021**, *16*, 1719–1723. [CrossRef] [PubMed]
2. West, S.W.; Clubb, J.; Torres-Ronda, L.; Howells, D.; Leng, E.; Vescovi, J.D. More than a Metric: How Training Load is Used in Elite Sport for Athlete Management. *Int. J. Sports Med.* **2021**, *42*, 300–306. [CrossRef] [PubMed]
3. Cerda-Kohler, H.; Aedo-Muñoz, E.; Solar-Altamirano, I. Lineamientos Unidad de Ciencias Aplicadas al Deporte, 1st ed.; Instituto Nacional de Deportes; 2022. Available online: https://www.researchgate.net/publication/359047294_Lineamientos_Unidad_de_Ciencias_Aplicadas_al_Deporte_2022-2028_El_que_el_como_y_el_por_que_hacemos_lo_que_hacemos (accessed on 28 August 2023).
4. McGuinness, A.; Malone, S.; Petrakos, G.; Collins, K. The Physical and Physiological Demands of Elite International Female Field Hockey Players During Competitive Match Play. *J. Strength Cond. Res.* **2019**, *33*, 3105–3113. [CrossRef] [PubMed]
5. Laursen, P.; Buchheit, M. *Science and Application of High-Intensity Interval Training: Solutions to the Programming Puzzle*, 1st ed.; Human Kinetics: Champaign, IL, USA, 2019; Available online: https://www.human-kinetics.co.uk/9781492552123/science-and-application-of-high-intensity-interval-training/ (accessed on 28 August 2023).
6. Cormie, P.; McGuigan, M.; Newton, R. Developing maximal neuromuscular power: Part 1—Biological basis of maximal power production. *Sports Med.* **2011**, *41*, 17–38. [CrossRef] [PubMed]
7. Spiteri, T.; Newton, R.U.; Binetti, M.; Hart, N.H.; Sheppard, J.M.; Nimphius, S. Mechanical determinants of faster change of direction and agility performance in basquetball athletes. *J. Strength Cond. Res.* **2015**, *29*, 2205–2214. [CrossRef]
8. Young, W.; Rayner, R.; Talpey, S. It's Time to Change Direction on Agility Research: A Call to Action. *Sports Med.* **2021**, *7*, 12. [CrossRef]
9. Brughelli, M.; Cronin, J.; Levin, G. Understanding Change of Direction Ability in Sport: A Review of Resistance Training Studies. *Sports Med.* **2008**, *38*, 1045–1060. [CrossRef]

10. Spencer, M.; Lawrence, S.; Rechichi, C.; Bishop, D.; Dawson, B.; Goodman, C. Time-motion analysis of elite field hockey, with special reference to repeated-sprint activity. *J. Sports Sci.* **2004**, *22*, 843–850. [CrossRef]
11. Spencer, M.; Bishop, D.; Dawson, B.; Goodman, C. Physiological and Metabolic Responses of Repeated-Sprint Activities Specific to Field-Based Team Sports. *Sports Med.* **2005**, *35*, 1025–1044. [CrossRef]
12. Gabbett, T. GPS Analisys of elite women's field hockey training and competition. *J. Strength Cond. Res.* **2010**, *25*, 1321–1324. [CrossRef]
13. McGuinness, A.; Malone, S.; Hughes, B.; Collins, K.; Passmore, D. Physical activity and physiological profiles of elite international female field hockey players across the quarters of competitive match play. *J. Strength Cond. Res.* **2019**, *33*, 2513–2522. [CrossRef] [PubMed]
14. Buchheit, M.; Samozino, P.; Glynn, J.; Ben, M.; Haddad, H.; Mendez-Villanueva, A.; Morin, J.B. Mechanical determinants of acceleration and maximal sprinting speed in highly trained young soccer players. *J. Sports Sci.* **2014**, *32*, 1906–1913. [CrossRef] [PubMed]
15. Jiménez-Reyes, P.; García-Ramos, A.; Cuadrado-Peñafiel, V. Differences in sprint mechanical force-velocity profile between trained soccer and futsal players. *Int. J. Sports Physiol. Perform.* **2018**, *14*, 478–485. [CrossRef]
16. Vanwanseele, B.; Aeles, J. Mechanical determinants of sprint performance in elite and sub-elite field hockey players. In Proceedings of the International Conference on Biomechanics in Sports, Tsukuba, Japan, 18–22 July 2016.
17. Samozino, P.; Rabita, G.; Dorel, S.; Slawinski, N.; Peyrot, N.; Saez de Villarreal, E.; Morin, J.B. A simple method for measuring power, force, velocity properties, and mechanical effectiveness in sprint running. *Scand. J. Med. Sci. Sports* **2016**, *26*, 648–658. [CrossRef] [PubMed]
18. Devismes, M.; Aeles, J.; Philips, J.; Vanwanseele, B. Sprint force-velocity profiles in soccer players: Impact of sex and playing level. *Sports Biomech.* **2021**, *20*, 947–957. [CrossRef]
19. Stavridis, I.; Smilios, I.; Tsopanidou, A.; Economou, T.; Paradisis, G. Differences in the Force Velocity Mechanical Profile and the Effectiveness of Force Application During Sprint-Acceleration Between Sprinters and Hurdlers. *Front. Sports Act. Living* **2019**, *1*, 26. [CrossRef]
20. Perez, J.; Guilhem, G.; Brocherie, F. Truncated Estimation of Skating Force-Velocity Profiling When Using High-Speed Video-Based Methods Compared to Radar-Derived Processing. *Front. Bioeng. Biotechnol.* **2021**, *9*, 661744. [CrossRef]
21. Stenroth, L.; Vartiainen, P.; Karjalainen, P. Force-velocity profiling in ice hockey skating: Reliability and validity of a simple, low-cost field method. *Sports Biomech.* **2020**, *22*, 874–889. [CrossRef]
22. Baena-Raya, A.; Soriano-Maldonado, A.; Conceição, F.; Jiménez-Reyes, P.; Rodríguez-Pérez, M. Association of the vertical and horizontal force-velocity profile and acceleration with change of direction ability in various sports. *Eur. J. Sport Sci.* **2021**, *21*, 1659–1667. [CrossRef]
23. Lythe, J.; Kilding, A.E. Physical demands and physiological responses during elite field hockey. *Int. J. Sports Med.* **2011**, *32*, 523–528. [CrossRef]
24. Buglione, A.; Ruscello, B.; Milia, R.; Migliaccio, G.M.; Granatelli, G.; D'Ottavio, S. Physical and Physiological demands of elite and sub-elite Field Hockey players. *Int. J. Perform. Anal. Sport* **2013**, *13*, 872–884. [CrossRef]
25. Harper, D.J.; Morin, J.B.; Carling, C.; Kiely, J. Measuring maximal horizontal deceleration ability using radar technology: Reliability and sensitivity of kinematic and kinetic variables. *Sports Biomech.* **2020**, *22*, 1192–1208. [CrossRef] [PubMed]
26. Harper, D.J.; Cohen, D.D.; Carling, C.; Kiely, J. Can countermovement jump neuromuscular performance qualities differentiate maximal horizontal deceleration ability in team sport athletes? *Sports* **2020**, *8*, 76. [CrossRef]
27. Mendiguchia, J.; Edouard, P.; Samozino, P.; Brughelli, M.; Cross, M.; Gill, N.; Morin, J.B. Field monitoring of sprinting power–force–velocity profile before, during and after hamstring injury: Two case reports. *J. Sports Sci.* **2016**, *34*, 535–541. [CrossRef] [PubMed]
28. Mendiguchia, J.; Samozino, P.; Martinez-Ruiz, E.; Schmikli, S.; Morin, J.B.; Mendez-Villanueva, A. Progression of mechanical properties during on-field sprint running after returning to sports from a hamstring muscle injury in soccer players. *Int. J. Sports Med.* **2014**, *35*, 690–695. [CrossRef] [PubMed]
29. Morin, J.B.; Samozino, P. Interpreting Power-Force-Velocity Profiles for Individualized and Specific Training. *Int. J. Sports Physiol. Perform.* **2016**, *11*, 267–272. [CrossRef]
30. Simperingham, K.; Cronin, J.; Ross, A. Advances in Sprint Acceleration Profiling for Field-Based Team-Sport Athletes: Utility, Reliability, Validity and Limitations. *Sports Med.* **2016**, *46*, 1619–1645. [CrossRef]
31. Jia, Y.; Guo, L.; Wang, X. Real-time control systems. In *Transportation Cyber-Physical Systems*; Elsevier: Amsterdam, The Netherlands, 2018; pp. 81–113. [CrossRef]
32. Jiménez-Reyes, P.; Samozino, P.; García-Ramos, A.; Cuadrado-Peñafiel, V.; Brughelli, M.; Morin, J.B. Relationship between vertical and horizontal force-velocity-power profiles in various sports and levels of practice. *PeerJ* **2018**, *6*, e5937. [CrossRef]
33. Castagna, C.; Impellizzeri, F.M.; Bizzini, M.; Weston, M.; Manzi, V. Applicability of a change of direction ability field test in soccer assistant referees. *J. Strength Cond. Res.* **2011**, *25*, 860–866. [CrossRef]
34. Morin, J.B.; Samozino, P.; Spreadsheet for Sprint Acceleration Force-Velocity-Power Profiling. Sport Science. 2017. Available online: https://jbmorin.net/2017/12/13/aspreadsheet-for-sprint-acceleration-force-velocity-powerprofiling/ (accessed on 5 November 2022).

35. Simperingham, K.D.; Cronin, J.B.; Pearson, S.N.; Ross, A. Reliability of horizontal force–velocity–power profiling during short sprint-running accelerations using radar technology. *Sports Biomech.* **2019**, *18*, 88–99. [CrossRef]
36. Vatcheva, K.; Lee, M.; McCormick, J.; Rahbar, M. Multicollinearity in Regression Analyses Conducted in Epidemiologic Studies. *Epidemiology* **2016**, *6*, 227. [CrossRef] [PubMed]
37. Shrestha, N. Detecting Multicollinearity in Regression Analysis. *Am. J. Appl. Math. Stat.* **2020**, *8*, 39–42. [CrossRef]
38. Eertink, J.; Heymans, M.; Zwezerijnen, G.; Zijlstra, J.; de Vet, H.; Boellaard, R. External validation: A simulation study to compare cross-validation versus holdout or external testing to assess the performance of clinical prediction models using PET data from DLBCL patients. *EJNMMI Res.* **2022**, *12*, 58. [CrossRef] [PubMed]
39. Geroldinger, A.; Lusa, L.; Nold, M.; Heinze, G. Leave-one-out cross-validation, penalization, and differential bias of some prediction model performance measures—A simulation study. *Diagn. Progn. Res.* **2023**, *7*, 9. [CrossRef] [PubMed]
40. Kipp, K.; Warmenhoven, J. Applications of regularized regression models in sports biomechanics research. *Sports Biomech.* **2022**, 1–19. [CrossRef]
41. Przednowek, K.; Iskra, J.; Maszczyk, A.; Nawrocka, M. Regression shrinkage and neural models in predicting the results of 400-metres hurdles races. *Biol. Sport* **2016**, *33*, 415–421. [CrossRef]
42. Sleivert, G.; Taingahue, M. The relationship between maximal jump-squat power and sprint acceleration in athletes. *Eur. J. Appl. Physiol.* **2004**, *91*, 46–52. [CrossRef]
43. Vescovi, J.D. Impact of maximum speed on sprint performance during high-level youth female field hockey matches: Female athletes in motion (FAiM) study. *Int. J. Sports Physiol. Perform.* **2014**, *9*, 621–626. [CrossRef]
44. González-Frutos, P.; Aguilar-Navarro, M.; Morencos, E.; Mallo, J.; Veiga, S. Relationships between strength and step frequency with fatigue index in repeated sprint ability. *Int. J. Environ. Res. Public Health* **2022**, *19*, 196. [CrossRef]
45. Samozino, P.; Peyrot, N.; Edouard, P.; Nagahara, R.; Jimenez-Reyes, P.; Vanwanseele, B.; Morin, J. Optimal mechanical force-velocity profile for sprint acceleration performance. *Scand. J. Med. Sci. Sports* **2022**, *32*, 559–575. [CrossRef]
46. Hicks, D.; Drummond, C.; Williams, K.; Van den Tillaar, R. Investigating Vertical and Horizontal Force-Velocity Profiles in Club-Level Field Hockey Athletes: Do Mechanical Characteristics Transfer Between Orientation of Movement? *Int. J. Strength Cond.* **2023**, *3*, 1–14. [CrossRef]
47. Haugen, T.; Breitschädel, F.; Seiler, S. Sprint mechanical variables in elite athletes: Are force-velocity profiles sport specific or individual? *PLoS ONE* **2019**, *14*, e215551. [CrossRef] [PubMed]
48. FIFA. *Pruebas de Condición Física para los Árbitros*; Ferecación Internacional de Fútbol Asociación (FIFA): Zürich, Switzerland, 2020.
49. Weston, M.; Helsen, W.; MacMahon, C.; Kirkendall, D. The Impact of Specific High-Intensity Training Sessions on Football Referees' Fitness Levels. *Am. J. Sports Med.* **2004**, *32*, 56–61. [CrossRef] [PubMed]
50. Harper, D.J.; Cohen, D.D.; Rhodes, D.; Carling, C.; Kiely, J. Drop jump neuromuscular performance qualities associated with maximal horizontal deceleration ability in team sport athletes. *Eur. J. Sport Sci.* **2022**, *22*, 1005–1016. [CrossRef] [PubMed]
51. Burgos-Jara, C.; Cerda-Kohler, H.; Aedo-Muñoz, E.; Miarka, B. Eccentric Resistance Training: A Methodological Proposal of Eccentric Muscle Exercise Classification Based on Exercise Complexity, Training Objectives, Methods, and Intensity. *Appl. Sci.* **2023**, *13*, 7969. [CrossRef]
52. McBurnie, A.J.; Harper, D.J.; Jones, P.A.; Dos'Santos, T. Deceleration Training in Team Sports: Another Potential 'Vaccine' for Sports-Related Injury? *Sports Med.* **2022**, *52*, 1–12. [CrossRef]
53. Brughelli, M.; Samozino, P. Mechanical Determinants of Sprinting Acceleration in Rugby Athletes. In Proceedings of the 23 Annual Congress of the European College of Sport Science, Dublin, Ireland, 4–7 July 2018; Available online: https://www.researchgate.net/publication/326252262 (accessed on 17 June 2023).
54. Morin, J.B.; Bourdin, M.; Edouard, P.; Peyrot, N.; Samozino, P.; Lacour, J.R. Mechanical determinants of 100-m sprint running performance. *Eur. J. Appl. Physiol.* **2012**, *112*, 3921–3930. [CrossRef]
55. Weyand, P.G.; Sandell, R.F.; Prime, D.N.; Bundle, M.W. The biological limits to running speed are imposed from the ground up. *J. Appl. Physiol.* **2010**, *108*, 950–961. [CrossRef]
56. Hunter, J.P.; Marshall, R.N.; McNair, P.J. Relationships between ground reaction force impulse and kinematics of sprint-running acceleration. *J. Appl. Biomech.* **2005**, *21*, 31–43. [CrossRef]
57. Morin, J.; Edouard, P.; Samozino, P. Technical ability of force application as a determinant factor of sprint performance. *Med. Sci. Sports Exerc.* **2011**, *43*, 1680–1688. [CrossRef]
58. Spiteri, T.; Cochrane, J.L.; Hart, N.H.; Haff, G.G.; Nimphius, S. Effect of strength on plant foot kinetics and kinematics during a change of direction task. *Eur. J. Sport Sci.* **2013**, *13*, 646–652. [CrossRef] [PubMed]
59. Nygaard, H.; Guldteig, H.; Van den Tillaar, R. Effect of Different Physical Training Forms on Change of Direction Ability: A Systematic Review and Meta-analysis. *Sports Med.* **2019**, *5*, 53. [CrossRef] [PubMed]
60. Zhang, Q.; Dellal, A.; Chamari, K.; Igonin, P.H.; Martin, C.; Hautier, C. The influence of short sprint performance, acceleration, and deceleration mechanical properties on change of direction ability in soccer players—A cross-sectional study. *Front. Physiol.* **2022**, *13*, 1027811. [CrossRef]
61. Harper, D.J.; McBurnie, A.J.; Santos, T.D.; Eriksrud, O.; Evans, M.; Cohen, D.D. Biomechanical and Neuromuscular Performance Requirements of Horizontal Deceleration: A Review with Implications for Random Intermittent Multi-Directional Sports. *Sports Med.* **2022**, *52*, 2321–2354. [CrossRef]

62. Van der Kruk, E.; Reijne, M. Accuracy of human motion capture systems for sport applications; state-of-the-art review. *Eur. J. Sport Sci.* **2018**, *18*, 806–819. [CrossRef] [PubMed]
63. Knudson, D. Letter to editor regarding 'the correlation of force-velocity-power relationship of a whole-body movement with 20 m and 60 m sprint'. *Sport Biomech.* **2021**. [CrossRef] [PubMed]
64. Haugen, T.; Seiler, S.; Sandbakk, Ø.; Tønnessen, E. The Training and Development of Elite Sprint Performance: An Integration of Scientific and Best Practice Literature. *Sports Med. Open* **2019**, *5*, 44. [CrossRef]

sensors

MDPI

Communication

Reliability of Maximal Strength and Peak Rate of Force Development in a Portable Nordic Hamstrings Exercise Device

Júlio A. Costa [1,*], Konstantinos Spyrou [1,2,3], António Sancho [4], Joana F. Reis [1,4,5] and João Brito [1]

1 Portugal Football School, Portuguese Football Federation (FPF), 1495-433 Cruz Quebrada, Portugal
2 UCAM Research Center for High Performance Sport, UCAM Universidad Católica de Murcia, 30107 Murcia, Spain
3 Facultad de Deporte, UCAM Universidad Católica de Murcia, 30107 Murcia, Spain
4 Faculdade de Motricidade Humana, Universidade de Lisboa, 1495-751 Lisboa, Portugal
5 Interdisciplinary Center for the Study of Human Performance (CIPER), Faculdade de Motricidade Humana, Universidade de Lisboa, 1495-751 Lisboa, Portugal
* Correspondence: jahdc@hotmail.com

Abstract: The Nordic hamstring exercise (NHE) is a very popular exercise used to improve eccentric strength and prevent injuries. The aim of this investigation was to assess the reliability of a portable dynamometer that measures maximal strength (MS) and rate of force development (RFD) during the NHE. Seventeen physically active participants (34.8 ± 4.1 years; n = 2 women and n = 15 men) participated. Measurements occurred on two different days separated by 48–72 h. Test–retest reliability was calculated for bilateral MS and RFD. No significant test–retest differences were observed in NHE (test–retest [95% CI, confidence interval]) for MS [−19.2 N (−67.8; 29.4); p = 0.42] and RFD [−70.4 N·s^{-1} (−178.4; 37.8); p = 0.19]. MS showed high reliability (intraclass correlation coefficient [ICC] [95% CI], =0.93 [0.80–0.97] and large within-subject correlation between test and retest [r = 0.88 (0.68; 0.95)]. RFD displayed good reliability [ICC = 0.76 (0.35; 0.91)] and moderate within-subject correlation between test and retest [r = 0.63 (0.22; 0.85)]. Bilateral MS and RFD displayed a coefficient of variation of 3.4% and 4.6%, respectively, between tests. The standard error of measurement and the minimal detectable change for MS was 44.6 arbitrary units (a.u.) and 123.6 a.u., and 104.6 a.u. and 290.0 a.u. for peak RFD. This study shows that MS and RFD can be measured for NHE using a portable dynamometer. However, not all exercises are suitable to apply to determine RFD, so caution must be taken when analyzing RFD during NHE.

Keywords: muscle strength; rehabilitation; groin; repeatability; reproducibility

Citation: Costa, J.A.; Spyrou, K.; Sancho, A.; Reis, J.F.; Brito, J. Reliability of Maximal Strength and Peak Rate of Force Development in a Portable Nordic Hamstrings Exercise Device. *Sensors* **2023**, *23*, 5452. https://doi.org/10.3390/s23125452

Academic Editors: Bruno Gonçalves, Bruno Emanuel Nogueira Figueira and Diogo Alexandre Martins Coutinho

Received: 12 April 2023
Revised: 6 June 2023
Accepted: 7 June 2023
Published: 9 June 2023

1. Introduction

Research into hamstring injuries has dramatically increased in the last two decades, because hamstring injuries are one of the most common injuries in high-speed running sports [1,2]. Specifically, one recent study conducted by Ekstrand et al. [1] reported that all hamstring injuries diagnosed in soccer in the 21-year study period have increased from 12% to 24%. Furthermore, the proportion of injury absence days caused by hamstring injuries increased from 10% to 20% [1]. Hamstring injuries are more likely to occur during running and sprinting, because the hamstring muscles experience the greatest amount of eccentric force during the late swing phase in the gait cycle [3], as the hip and knee muscles during late swing phase demonstrated the most dramatic increase in biomechanical load (i.e., torques, net powers, and work done) when running speed progressed [4–6]. Furthermore, hamstring injury depends on many factors [7]. Specifically, hamstring eccentric and concentric strength, lumbopelvic and knee stability, lower-limb stiffness, and insufficient sprint exposure may increase the likelihood of a hamstring injury occurring [7]. Given that, eccentric knee flexor muscle strength is one of the fundamental metrics to prevent injuries [8–10] and consequently increase an athlete's specific performance (e.g., acceleration and high-speed running) [8,11,12].

The Nordic hamstring exercise (NHE) is one of the most common eccentric exercises used in sports and is currently used as a major injury prevention strategy [12]. The exercise instruction is the following: the athlete is asked to perform eccentric knee flexor maximal strength (MS) in a high kneeling position with the ankles fixed either by a partner or by a stationary object. From this position, the athlete inclines the torso, maintaining neutral hip alignment, for as far as possible and then uses the arms to contact the ground in front when the hamstrings can no longer control the downward movement. The NHE intervention seems to increase the fascicle strength by increasing the number of sarcomeres in series within the muscle fibers, and there are potential changes in the distribution through electromyography of the three biarticular of the hamstring muscles (i.e., biceps femoris long head, semitendinosus, and the semimembranosus) [13–15].

Regarding previous research, Lodge et al. [16] found high test–retest reliability, ICC 0.91 (CI, 0.76–0.96) and 0.91 (CI, 0.78–0.96) for left and right eccentric knee flexor muscle strength peak forces, respectively, using an eccentric hamstring strength measurement device similar to the portable dynamometer used in the current study compared to an isokinetic dynamometer. Furthermore, similar results were found in inter-rater reliability and correlations between isometric and eccentric knee extension and flexion strength using a hand-held dynamometry and isokinetic test for knee flexion extension. Consequently, it is vital to highlight that eccentric knee flexor muscle strength devices have already been validated, and the aim of this study was to evaluate the test–retest reliability of a portable dynamometer. To the authors' knowledge, this is the first study that has measured reliability (i.e., test–retest) of the rate of force development (RFD) during knee flexor strength testing.

To date, the gold standard measure for the evaluation of eccentric knee flexor strength is isokinetic dynamometry [17]. However, isokinetic dynamometers are characterized by a lack of portability, high cost and time consumption, and their daily use might be practically difficult. Considering that a great number of devices that use load cell dynamometers have become popular field-based methods to monitor individual eccentric knee flexor strength during a NHE [9,17–19], therefore, the aim of this study was to evaluate the test–retest reliability of eccentric knee flexor MS and peak RFD during NHE using a portable dynamometer. The leading hypothesis of the current study was that the current portable dynamometer provides reliable data of eccentric knee flexor MS and RFD, and the study was designed to answer the main research question declared above.

2. Materials and Methods

2.1. Participants

Seventeen healthy and physically active adult subjects (n = 15 men and n = 2 women) who engaged in more than 3 h of physical exercise per week, who were injury free in the lower limbs, and had no pain or illness in the past 3 weeks before starting the study volunteered to take part in this study. Table 1 reports the characteristics of the participants. The experimental design and potential risks of the study were explained to the participants and written informed consent was provided. The study was approved by the Ethics Committee of the Portugal Football School, Portuguese Football Federation (CEPFS 12.2021).

Table 1. Participants' characteristics (n = 17).

	Total (n = 17)
Age (years)	34.8 ± 4.1
Body mass (kg)	78.5 ± 16.2
Height (m)	1.8 ± 0.1
BMI (kg/m^2)	24.1 ± 3.6

Values are expressed in mean ± standard deviation. BMI, body max index.

2.2. Procedures

The same researcher recorded the test–retest NHE performance data at three distinct sessions on different days. Firstly, the participants performed a familiarization session that included the same warm-up, order, and exercises as the evaluation sessions. Approximately 7 days later, the participants performed the first test session, and the retest session was conducted within 48–72 h from the end of the first test session. All sessions were conducted at the same time of the day (i.e., during morning or afternoon). Participants were asked to not perform any vigorous lower-limb exercises in the 24 h before each testing session [20].

In the first session, the participants performed a warm-up that consisted of 7 min on a bicycle ergometer at a pedaling cadence of 75–80 rpm, 2 sets of 12 reps of half-squats, standing toe raises, and hip bridge [21]. According to recent literature [21], participants were positioned in a kneeling position over the padded board, ankles held under lockable braces (fixed atop the uniaxial load cells), with the lateral malleolus aligned with the edge of the board and arms across the chest, using the portable dynamometer (Figure 1).

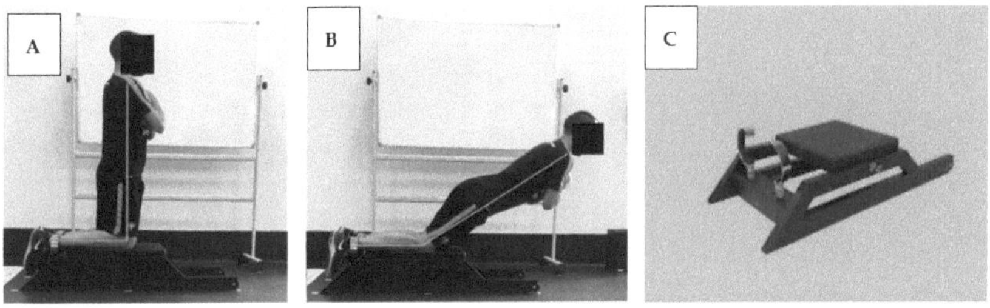

Figure 1. Testing set-up. (**A**) starting positioning; (**B**) participant leaned forward slowly and as controlled as possible (eccentric phase only); (**C**) portable dynamometer used in this study.

The dynamometer Smart Nordic Trainer (Neuroexcellence®; S-2A INOX, Porto, Portugal) has two load cells (one on each hook to measure the force applied by each leg). Each cell has a maximum capacity of 4903 N (~500 kg). When starting the movement, the reading of the cell is correct, but in the middle of the exercise, the hook has a rotational movement of about 5 to 8 degrees, which is intentional, which is the adaptation of the hook to the athlete's exercise, which can vary the angle from athlete to athlete. The manufacturers considered this read error to be negligible. A load of 100 kg with a hook rotated by 8 degrees corresponds to an error of ±1 kg. The cell reading is 100 g. Model SENSOCAR® S-2A INOX has a repeatability error <0.02% F.E, sensitivity 2.0 mV/V ± 0.1%, zero offset < 1% F.E, combined maximum error < 0.02% F.E., Fluence 30 min (creep) < 0.02%. The metrics were calculated according to the manufacturer as follows:

N: number of recorded samples; F: Force list; t: Timestamp list

Maximal Force: MaxValue = max(F)

$$\text{Maximal Force}: \text{MaxValue} = \max(F)$$
$$\text{Peak RFD} = \boldsymbol{max}(\{f(x) : x \in [N..1]\})$$
$$f(x) = \frac{Fx - Fx - n}{tx - tx - n}$$

where n is the closest index and $t_x - t_{x-n}$ is equal to the Time Interval RFD. Note: the default value of the Time Interval RFD is 0.05 s.

During the familiarization, and during test and retest sessions, the participants performed 3 maximal trials of eccentric NHE repetitions, where participants leaned forward in a slow, controlled manner for as long as possible, during the eccentric phase. The movement onset was determined by counting down from three to one (information given by the software), and then the participant started performing the NHE. Then, the participants passively returned to the starting position, in order to repeat the following repetitions. According to recent literature [21], the maximal NHE trials were separated by a standardized

1-min rest period to allow for recovery and to avoid fatigue. If participants increased their performance in all three trials, one or two additional NHE repetitions were performed [21]. Authorized feedback from the investigator was used to motivate the subjects. Trials were only regarded as successful if the participants held trunk and hips in a neutral position during the NHE repetitions. Participants controlled the movement until they lost control and stopped dealing with it. No additional loading was used. According to the ANHEQ criteria [22], the total score for NHE quality was 8 points, which is considered "good". Specifically, ANHEQ criteria are the following: (1) Rigid fixation: 2 points; (2) Knee position: 1 point; (3) Kneeling height: 1 point; (4) Separate familiarization: 1 point; (5) Diagnostic tools: 1 point; (6) Feedback of target movement speed: 0 points—we only provided feedback to the participant to perform as slowly as possible; (7) Consequences of impaired technique: 1 point; (8) Presentation of NHE performance variables: 1 point. Bilateral MS and RFD were considered for analysis. All data were recorded with corporative data acquisition software (NexSo v1.0.0., Porto, Portugal). Data were collected through the Bluetooth BLE communication protocol at 180 Hz. The tests were performed in a gym facility.

2.3. Statistical Analyses

Sample distribution was tested using the Shapiro–Wilk test for MS and peak RFD variables. Variables are presented as mean with the 95% confidence interval (CI).

Linear mixed model analysis was performed to examine differences in the MS and peak RFD during test–retest.

To estimate the test–retest reliability of the NHE, intraclass correlation coefficients (ICC) [23] and the two-way random effects model of the measurements with 95% CI was used. The ICC were classified in the following manner: >0.90, high reliability; 0.80–0.89, good reliability; between 0.70 and 0.79, fair reliability; and values <0.69, poor reliability [24]. Further, within-subject variation was determined using typical error expressed as a coefficient of variation (CV) [25].

The standard error of measurement [25] and the minimal detectable change (MDC) were calculated to analyze the variability of the participants' performances. For this analysis, the following formulas were used to calculate the SEM and MDC [25].

$$SEM = SD \times \sqrt{(1 - ICC)}$$

$$MDC = SEM \times \sqrt{2} \times 1.96$$

We tested the within-subject correlations (r, 95% CI) [26] between test and retest for MS and peak RFD variables. We qualitatively interpreted the magnitudes of correlation using the following criteria: trivial ($r \leq 0.1$), small (r = 0.1–0.3), moderate (r = 0.3–0.5), large (r = 0.5–0.7), very large (r = 0.7–0.9), and almost perfect ($r \geq 0.9$) [27].

Most of the statistical analyses were conducted using SPSS software (version 27.0.1, SPSS Inc., Chicago, IL, USA), except for within-subject correlation for which a rmcorr package in R statistical software (version 3.4.1, R Foundation for Statistical Computing, Vienna, Austria) was used.

3. Results

Values of bilateral absolute MS and RFD variables during the NHE in healthy and physical activity adults are presented in Figure 2.

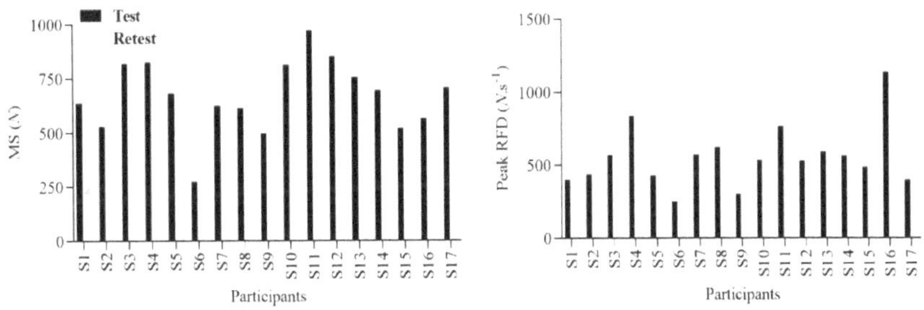

Figure 2. Descriptive test–retest individual data for MS and peak RFD during NHE in healthy and physically active adults (n = 17).

During the familiarization session, MS was 669.4 N (581.1; 734.3) and peak RFD was 543.8 N·s^{-1} (428.7; 654.9). No significant test–retest differences were observed in NHE performance for MS and RFD variables (Table 2).

Table 2. Descriptive and test–retest differences data for bilateral MS, relative MS, and peak RFD, during NHE in healthy and physically active adults (n = 17).

	Test	Retest	Δ (Test–Retest)	p
Bilateral MS (N)	669.8 (583.8; 755.9)	689 (588.3; 789.7)	−19.2 (−67.8; 29.4)	$p = 0.42$
Relative MS (N/Kg)	8.6 (7.7; 8.9)	8.7 (7.8; 9.6)	−0.1 (−0.8; 0.5)	$p = 0.63$
Peak RFD (N·s^{-1})	554.9 (446.6; 663.1)	625.3 (488.6; 761.9)	−70.4 (−178.4; 37.8)	$p = 0.19$

Values are expressed in mean (95% CI).

MS showed high reliability (ICC = 0.93 [0.80–0.97]) and large within-subject correlation between test and retest [r = 0.88 (0.68; 0.95)] (Figure 3). Peak RFD demonstrated good reliability [ICC = 0.76 (0.35; 0.91)] and moderate within-subject correlation between test and retest [r = 0.63 (0.22; 0.85)]. The MS and peak RFD presented CV values of 3.4% and 4.6%, respectively, between test and retest. The standard error of measurement and the MDC for MS was 44.6 arbitrary units (a.u.) and 123.6 a.u., and for peak RFD was 104.6 a.u. and 290.0 a.u.

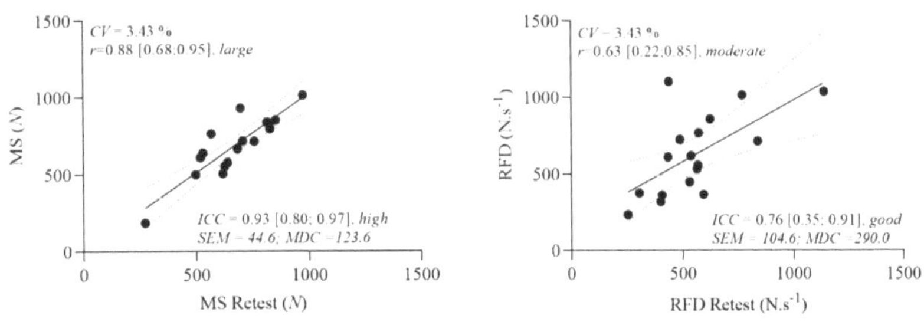

Figure 3. Test–retest reliability and a within-subject correlation was calculated for MS and peak RFD during NHE in healthy and physically active adults (n = 17). ICC, intraclass correlation coefficient [95%CI]; CV, coefficient of variation; SEM, standard error of measurement; MDC, minimal detectable change; N, Newton.

Figures 4 and 5 presents the force and time profile of the test–retest of a representative participant and the RFD in one repetition. Figure 6 depicts the force time profile of the first session of four randomly chosen participants in order to represent their different completions of the exercise.

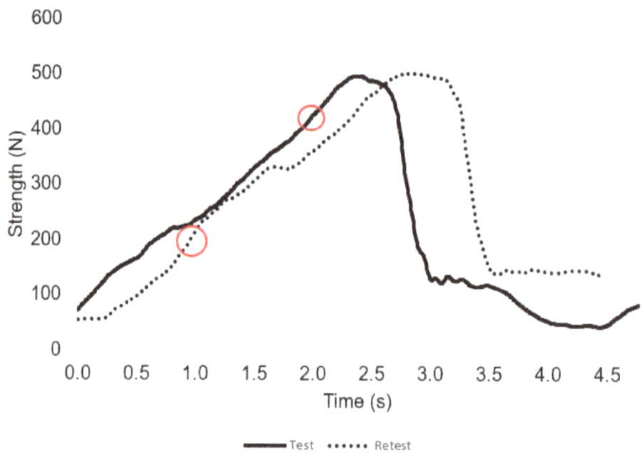

Figure 4. Test–retest individual data between strength (N) and time (ms) of a representative subject. Red circles symbolize the moment where the RFD was higher during NHE.

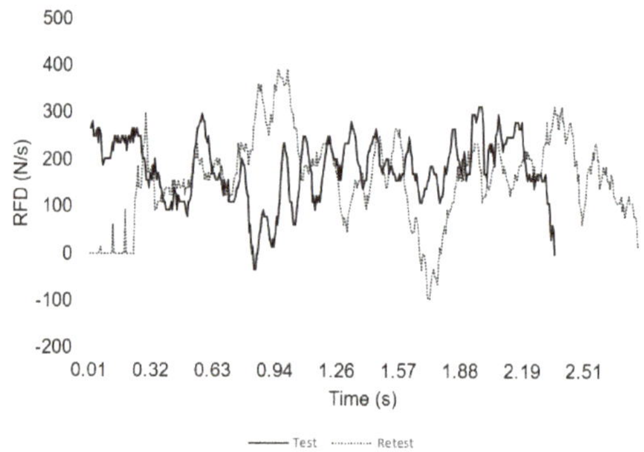

Figure 5. Test–retest individual data between RFD (N/s) and time (ms) of a representative subject.

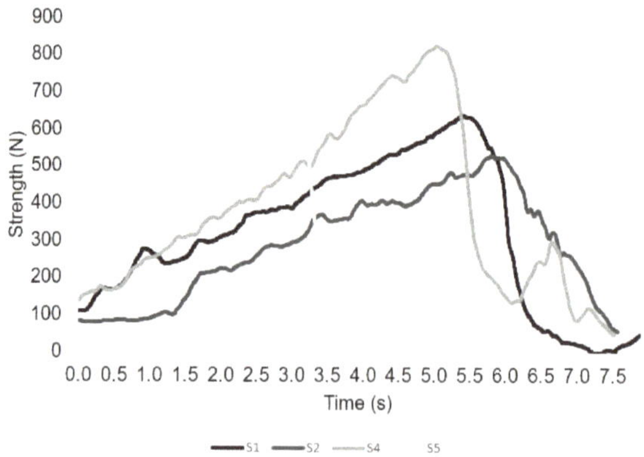

Figure 6. Example for different individual subjects (S1, S2, S3, S4, and S5) performing a NHE.

4. Discussion

The aim of this study was to evaluate the test–retest reliability of the bilateral eccentric knee muscle flexor MS and RFD during NHE. The main findings were the following: (1) no significant test–retest differences were observed in NHE for MS and RFD; (2) MS showed *high* reliability and *large* within-subject correlation between test and retest;

(3) RFD displayed *good* reliability and *moderate* within-subject correlation between test and retest; and (4) MS and RFD presented CV values of 3.4% and 4.6%, respectively, between test and retest. This study shows that MS and RFD can be measured for NHE using a portable dynamometer.

Regarding peak absolute strength, NHE showed *high* reliability (ICC = 0.93 and CV = 3.4%) and a *large* within-subject correlation between test and retest (r = 0.88) (Figure 3). The current results for test–retest reliability of the HSE device is in line with previous studies [9,17,19,28,29]. For example, Lodge et al. [17] found *high* test–retest reliability, ICC 0.91 (CI, 0.76–0.96), and 0.91 (CI, 0.78–0.96) for left and right eccentric knee flexor muscle strength peak forces, respectively, using an eccentric hamstring strength measurement device similar to the portable dynamometer used in the current study compared to an isokinetic dynamometer. Moreover, similar results showed an inter-rater reliability and correlations between the isometric and eccentric knee extension and flexion strengths using hand-held dynamometry and an isokinetic test for knee flexion extension of athletic participants. Therefore, it is vital to highlight that eccentric knee flexor muscle strength devices are already validated and the aim of this study was to evaluate the reliability of test–retest of a portable dynamometer. In a practical application, the current portable device can be used to evaluate and train eccentric flexor muscle strength on a daily basis.

Considering RFD showed *good* reliability (ICC = 0.76; 4.6%) and *moderate* within-subject correlation between test and retest (r = 0.63) (Figure 3). To the authors' knowledge, this is the first study measuring the RFD reliability (i.e., test–retest) during a knee flexor strength test. Compared with the assessment of RFD for the hip muscles (i.e., hip adductor, flexor, and external rotator), RFD for the hamstrings can be measured with confidence (i.e., ICC > 0.70 and standard error < 10%) [30]. Considering the *moderate* within-subject correlation between test and retest, it is important to highlight that RFD assessments might be challenging and need more time for familiarization with the test [31]; also, it is important to highlight that the NHE is not performed in a maximal isometric contraction, as it is in traditional RFD evaluations. Therefore, due to the controlled and slow movement in the NHE, the highest RFD may not occur at the beginning of the exercise. Therefore, RFD is acceptable to evaluate by the hamstring strength portable device, but it should be conducted with caution and familiarization. In an applied setting, the RFD value can be recorded from the portable dynamometer.

This study is limited by the NHE itself, as factors such as lack of control on the velocity of the movement, the intervention from other muscle groups, such as the lumbo-pelvic zone, and the determination of the "optimal" angle peak torque of the knee flexor muscle group, which would be useful when targeting strength improvements at a specific joint angle (that could be measured by the gold standard measurement such as an isokinetic dynamometer). Furthermore, regarding Assessing the NHE quality [21] scale knee position is a key component of NHE execution as, on a rigid surface, the pressure on the knees may cause an uncomfortable feeling and pain. Additionally, the RFD metric, even with *good* reliability [ICC = 0.76 (0.35; 0.91)] and *moderate* within-subject correlation between test and retest [r = 0.63 (0.22; 0.85)], should be used cautiously as it is a controlled and slow movement. The current study was designed to examine the test–retest reliability, considering essential to use on a daily basis, avoiding misrepresentation of changes in strength and minimizing the error of measurement. Further research about the validation of the current portable dynamometer when compared to gold standard measurements, such as isokinetic dynamometers or other similar portable devices that have already been validated, is warranted. Lastly, more investigation is warranted regarding NHE variations with rapid muscle activation, i.e., reactively bouncing and decelerating exercises which elicit much higher peak moments than the standard NHE [6].

5. Conclusions

In conclusion, the current device presented no significant test–retest differences during the NHE for MS and RFD. Furthermore, MS and RFD variables showed *good–high* reliability and *moderate–large* within-subject correlation between test and retest, respectively. Lastly, MS and RFD showed CV values of 3.4% and 4.6%, respectively, between test and retest. This study shows that MS can be measured during the NHE using a portable dynamometer, but this should be performed with caution and with previous familiarization due to the slow and controlled movement of the NHE that may not favor the attainment of the highest RFD at the beginning of exercise.

Author Contributions: Conceptualization, J.A.C. and J.B.; methodology, J.A.C. and J.B.; software, J.A.C. and A.S.; validation, J.B.; formal analysis, J.A.C. and A.S.; investigation, J.A.C., K.S., A.S. and J.B.; resources, J.B.; data curation, J.A.C. and A.S.; writing—original draft preparation, J.A.C. and K.S.; writing—review and editing, J.A.C., K.S., A.S., J.F.R. and J.B.; visualization, J.B.; supervision, J.B. All authors have read and agreed to the published version of the manuscript.

Funding: This research received no external funding.

Institutional Review Board Statement: The study was approved by the Ethics Committee of the Portugal Football School, Portuguese Football Federation (CEPFS 12.2021).

Informed Consent Statement: The study design was carefully explained to the subjects, and written informed consent was obtained.

Data Availability Statement: Data may be available from the Data Protection Office, Portuguese Football Federation (Data Access contact via e-mail: dpo@fpf.pt) for researchers who meet the criteria to access confidential data.

Acknowledgments: The authors would like to thank the subjects for their participation and cooperation during the study.

Conflicts of Interest: The authors declare no conflict of interest.

References

1. Ekstrand, J.; Bengtsson, H.; Walden, M.; Davison, M.; Khan, K.M.; Hagglund, M. Hamstring injury rates have increased during recent seasons and now constitute 24% of all injuries in men's professional football: The UEFA Elite Club Injury Study from 2001/02 to 2021/22. *Br. J. Sport. Med.* **2022**, *57*, 292–298. [CrossRef] [PubMed]
2. Hagglund, M.; Walden, M.; Ekstrand, J. Injury recurrence is lower at the highest professional football level than at national and amateur levels: Does sports medicine and sports physiotherapy deliver? *Br. J. Sport. Med.* **2016**, *50*, 751–758. [CrossRef] [PubMed]
3. Liu, H.; Garrett, W.E.; Moorman, C.T.; Yu, B. Injury rate, mechanism, and risk factors of hamstring strain injuries in sports: A review of the literature. *J. Sport Health Sci.* **2012**, *1*, 92–101. [CrossRef]
4. Schache, A.G.; Blanch, P.D.; Dorn, T.W.; Brown, N.A.; Rosemond, D.; Pandy, M.G. Effect of running speed on lower limb joint kinetics. *Med. Sci. Sport. Exerc.* **2011**, *43*, 1260–1271. [CrossRef]
5. Chumanov, E.S.; Heiderscheit, B.C.; Thelen, D.G. The effect of speed and influence of individual muscles on hamstring mechanics during the swing phase of sprinting. *J. Biomech.* **2007**, *40*, 3555–3562. [CrossRef]
6. Alt, T.; Roos, T.; Nolte, K.; Modenbach, D.; Knicker, A.J.; Jaitner, T. Modulating the Nordic Hamstring Exercise from 'zero to hero'—A Stepwise progression explored in a high-performance athlete. *J. Athl. Train.* 2022; *online ahead of print*. [CrossRef]
7. Huygaerts, S.; Cos, F.; Cohen, D.D.; Calleja-González, J.; Guitart, M.; Blazevich, A.J.; Alcaraz, P.E. Mechanisms of Hamstring Strain Injury: Interactions between Fatigue, Muscle Activation and Function. *Sports* **2020**, *8*, 65. [CrossRef]
8. Alt, T.; Severin, J.; Komnik, I.; Nodler, Y.T.; Benker, R.; Knicker, A.J.; Brüggemann, G.-P.; Strüder, H.K. Nordic Hamstring Exercise training induces improved lower-lim swing phase mechanics and sustained strength preservation in sprinters. *Scand. J. Med. Sci. Sport.* **2021**, *31*, 826–838. [CrossRef]
9. Opar, D.A.; Piatkowski, T.; Williams, M.D.; Shield, A.J. A novel device using the Nordic hamstring exercise to assess eccentric knee flexor strength: A reliability and retrospective injury study. *J. Orthop. Sport. Phys. Ther.* **2013**, *43*, 636–640. [CrossRef]
10. van Dyk, N.; Behan, F.P.; Whiteley, R. Including the Nordic hamstring exercise in injury prevention programmes halves the rate of hamstring injuries: A systematic review and meta-analysis of 8459 athletes. *Br. J. Sport. Med.* **2019**, *53*, 1362–1370. [CrossRef]
11. Siddle, J.; Greig, M.; Weaver, K.; Page, R.M.; Harper, D.; Brogden, C.M. Acute adaptations and subsequent preservation of strength and speed measures following a Nordic hamstring curl intervention: A randomised controlled trial. *J. Sport. Sci.* **2019**, *37*, 911–920. [CrossRef]
12. Ishøi, L.; Hölmich, P.; Aagaard, P.; Thorborg, K.; Bandholm, T.; Serner, A. Effects of the Nordic Hamstring exercise on sprint capacity in male football players: A randomized controlled trial. *J. Sport. Sci.* **2018**, *36*, 1663–1672. [CrossRef] [PubMed]
13. Loturco, I.; Freitas, T.T.; Alcaraz, P.E.; Kobal, R.; Hartmann Nunes, R.F.; Weldon, A.; Pereira, L.A. Practices of strength and conditioning coaches in Brazilian elite soccer. *Biol. Sport* **2022**, *39*, 779–791. [CrossRef] [PubMed]

14. Brockett, C.L.; Morgan, D.L.; Proske, U. Human hamstring muscles adapt to eccentric exercise by changing optimum length. *Med. Sci. Sport. Exerc.* **2001**, *33*, 783–790. [CrossRef]
15. Gérard, R.; Gojon, L.; Decleve, P.; Van Cant, J. The Effects of Eccentric Training on Biceps Femoris Architecture and Strength: A Systematic Review With Meta-Analysis. *J. Athl. Train.* **2020**, *55*, 501–514. [CrossRef] [PubMed]
16. Suskens, J.J.M.; Reurink, G.; Tol, J.L.; Kerkhoffs, G.M.M.J.; Goedhart, E.A.; Maas, H.; van Dieën, J.H. Activity Distribution Among the Hamstring Muscles During the Nordic Hamstring Exercise: A Multichannel Surface Electromyography Study. *J. Appl. Biomech.* **2023**, *39*, 69–79. [CrossRef]
17. Lodge, C.; Tobin, D.; O'Rourke, B.; Thorborg, K. Reliability and Validity of a New Eccentric Hamstring Strength Measurement Device. *Arch. Rehabil. Res. Clin. Transl.* **2020**, *2*, 100034. [CrossRef]
18. Aagaard, P.; Simonsen, E.B.; Magnusson, S.P.; Larsson, B.; Dyhre-Poulsen, P. A new concept for isokinetic hamstring: Quadriceps muscle strength ratio. *Am. J. Sport. Med.* **1998**, *26*, 231–237. [CrossRef]
19. Whiteley, R.; Jacobsen, P.; Prior, S.; Skazalski, C.; Otten, R.; Johnson, A. Correlation of isokinetic and novel hand-held dynamometry measures of knee flexion and extension strength testing. *J. Sci. Med. Sport* **2012**, *15*, 444–450. [CrossRef]
20. Gonçalves, B.A.M.; Mesquita, R.N.O.; Tavares, F.; Brito, J.; Correia, P.; Santos, P.; Mil-Homens, P. A New Portable Device to Reliably Measure Maximal Strength and Rate of Force Development of Hip Adduction and Abduction. *J. Strength Cond. Res.* **2022**, *36*, 2465–2471. [CrossRef]
21. Augustsson, J.; Augustsson, S.R. Development of a Novel Nordic Hamstring Exercise Performance Test Device: A Reliability and Intervention Study. *Sports* **2022**, *10*, 26. [CrossRef]
22. Alt, T.; Schmidt, M. The ANHEQ Evaluation Criteria: Introducing Reliable Rating Scales for Assessing Nordic Hamstring Exercise Quality. *Sport. Med.-Open* **2021**, *7*, 91. [CrossRef] [PubMed]
23. Shrout, P.E.; Fleiss, J.L. Intraclass correlations: Uses in assessing rater reliability. *Psychol. Bull.* **1979**, *86*, 420–428. [CrossRef] [PubMed]
24. Hopkins, W.G. Measures of reliability in sports medicine and science. *Sport. Med.* **2000**, *30*, 1–15. [CrossRef] [PubMed]
25. Haley, S.M.; Fragala-Pinkham, M.A. Interpreting change scores of tests and measures used in physical therapy. *Phys. Ther.* **2006**, *86*, 735–743. [CrossRef] [PubMed]
26. Bland, J.M.; Altman, D.G. Calculating correlation coefficients with repeated observations: Part 1—Correlation within subjects. *BMJ* **1995**, *310*, 446. [CrossRef]
27. Hopkins, W.G.; Marshall, S.W.; Batterham, A.M.; Hanin, J. Progressive statistics for studies in sports medicine and exercise science. *Med. Sci. Sport. Exerc.* **2009**, *41*, 3–13. [CrossRef]
28. Impellizzeri, F.M.; Bizzini, M.; Rampinini, E.; Cereda, F.; Maffiuletti, N.A. Reliability of isokinetic strength imbalance ratios measured using the Cybex NORM dynamometer. *Clin. Physiol. Funct. Imaging* **2008**, *28*, 113–119. [CrossRef]
29. Li, R.C.; Wu, Y.; Maffulli, N.; Chan, K.M.; Chan, J.L. Eccentric and concentric isokinetic knee flexion and extension: A reliability study using the Cybex 6000 dynamometer. *Br. J. Sport. Med.* **1996**, *30*, 156–160. [CrossRef]
30. Casartelli, N.C.; Lepers, R.; Maffiuletti, N.A. Assessment of the rate of force development scaling factor for the hip muscles. *Muscle Nerve* **2014**, *50*, 932–938. [CrossRef]
31. Maffiuletti, N.A.; Aagaard, P.; Blazevich, A.J.; Folland, J.; Tillin, N.; Duchateau, J. Rate of force development: Physiological and methodological considerations. *Eur. J. Appl. Physiol.* **2016**, *116*, 1091–1116. [CrossRef]

sensors

MDPI

Article

Relationship between Repeated Sprint Ability, Countermovement Jump and Thermography in Elite Football Players

Carlos Majano [1], Jorge García-Unanue [1,*], Antonio Hernandez-Martin [1], Javier Sánchez-Sánchez [2], Leonor Gallardo [1] and Jose Luis Felipe [1]

1 IGOID Research Group, Physical Activity and Sport Sciences Department, University of Castilla-La Mancha, 45071 Toledo, Spain
2 School of Sport Sciences, Universidad Europea de Madrid, 28670 Villaviciosa de Odón, Spain
* Correspondence: jorge.garciaunanue@uclm.es

Abstract: Football is a very demanding sport which requires players to exert maximum effort, producing fatigue and eventually injuries. Thermography can be used to detect fatigue and prevent its consequences through thermal asymmetries in the bilateral body areas; however, its adequacy for elite footballers has not been widely studied. Therefore, the objective of the present investigation was to determine the suitability of thermography to detect fatigue in male football players. For this reason, twenty participants were gathered into a pair of subgroups (low [<0.2 °C] vs. high thermal asymmetry [≥0.2 °C]) based on a thermography session of the lower limbs (thighs, calves, and hamstrings). After the thermography session, players performed CMJs before and after an RSA test (6 × 30 m/20″). A mixed two-way analysis of variance and Bonferroni post hoc pairwise comparisons were undertaken to analyse the results. No significant differences ($p > 0.05$) were found in any of the RSA test variables between low and high thermal asymmetry groups for thighs and calves. On the other hand, the low thermal asymmetry hamstring group reported a smaller percentage difference in sprints for the first sprint (%Diff) and a larger percentage difference in sprints two and three with respect to the best sprint (%Best). For CMJs, the low thermal asymmetry hamstring group reported significantly higher values post-RSA test, indicating better performance. Accordingly, thermography can provide information about performance in CMJ and RSA tests through hamstring asymmetries over 0.2 °C. Meanwhile, larger asymmetries than 0.2 °C in calves and thighs do not seem to be related to performance in these tests; therefore, coaches should consider if it is optimal to align players with high hamstring asymmetries.

Keywords: asymmetries; fatigue; temperature

Citation: Majano, C.; García-Unanue, J.; Hernandez-Martin, A.; Sánchez-Sánchez, J.; Gallardo, L.; Felipe, J.L. Relationship between Repeated Sprint Ability, Countermovement Jump and Thermography in Elite Football Players. *Sensors* **2023**, *23*, 631. https://doi.org/10.3390/s23020631

Academic Editors: Bruno Gonçalves, Bruno Emanuel Nogueira Figueira and Diogo Alexandre Martins Coutinho

Received: 14 November 2022
Revised: 27 December 2022
Accepted: 3 January 2023
Published: 5 January 2023

1. Introduction

Injuries and performance are important factors in football, which is a sport characterized by continuous changes in activity, alternating high-intensity actions with short resting periods [1]. In addition, the game demands tackles, shots, jumps, accelerations, decelerations, changes of direction and dribbles, making the sport highly physiologically demanding [2]. This is why muscle injuries of the lower extremities occur frequently in elite and amateur football players during matches [3]; however, only 5% of these injuries occur as a consequence of faults and contacts, which means that it may be possible to prevent these injuries to some extent [4]. A congested calendar of matches and training can induce fatigue, understood as several alterations in the central nervous system that produce an inability to maintain the required level of strength or an inability to complete a physical task that was previously achievable [5]. In terms of football, fatigue can appear because some players can be unable to assimilate matches and training loads, increasing the risk of injury and underperformance [6]. In this sense, coaches and researchers agree about the importance of managing fatigue to achieve optimal muscular adaptations, increase progress, prevent overtraining, decrease injury risk and improve performance [7]. However, very few studies have actively investigated how to measure fatigue and adjust training accordingly.

Fatigue can be measured through internal and external loads [5]. In football, there are several internal load methods that are very popular to measure fatigue and response to exercise; however, most of them present some disadvantages, so they should be combined with other methods. Heart rate measurements are the most common, but they are imprecise in high-intensity actions and sprints, which are crucial aspects of football [8]. Blood and bone markers are more precise, but they are difficult to extract on a weekly basis (price and time), and some players can feel uncomfortable being exposed continuously to them. Therefore, extracting them whenever the physical status of the player needs to be known can be unsustainable during regular-season training [9]. Questionnaires are very simple to administer but particularly subjective, introducing a high risk of bias [10].

Conversely, there are some physical tests that can help to check the current physical condition of the athletes through external loads. It has been evidenced that velocity loss is a reliable marker of neuromuscular fatigue, which is described as a decrease in the physical and mental functions [11]. This is why most of the scientific literature has used the repeated sprint ability (RSA) test to measure fatigue, especially in team sports such as football; this is because RSA and the ability to exercise at high intensity are key capacities for optimal performance, and a decreased sprint repetition capacity is a good indicator of fatigue in sports [12]. However, this method can lack precision in some situations, as the scientific literature has found that sprint performance can be maintained in situations of neuromuscular fatigue after match play [13]. This is why some coaches incorporate a stretching–shortening movement, such as a countermovement jump (CMJ) to detect fatigue, as it can highlight impairments in jump performance between two events [14]; a good relationship between sprint ability and CMJ capacity has been verified [14].

Therefore, scientific evidence has shown CMJ performance to be an objective marker of fatigue and supercompensation, as neuromuscular fatigue has been associated with a decrease in the average CMJ height [15], causing this method to be used frequently to assess neuromuscular fatigue [16]. However, when using RSA to assess neuromuscular fatigue, the number of sprints induces great variability between athletes [17]. In addition, when using CMJs, it is important to consider that the same fatiguing stimuli can elicit different effects between individuals, and sometimes athletes can adjust their jumping strategy to maintain their performance [18]. Moreover, players' need to exert maximal effort to measure fatigue is not optimal, as this effort increases injury risk. Hence, the use of other technologies which can provide information and help coaches without the need to perform a physical test, can be considered. In this sense, one option is infrared thermography (IRT), which is a non-radiating, contact-free, safe and non-invasive technology that monitors physiological variables through the control of the skin temperature [19].

This technology gauges the correlation between muscle activation and skin temperature, as muscle and skin temperature are directly correlated. The information this technology provided is based on the hyperthermic and hypothermic responses of the skin [20], due to the fact that thermography is a reliable method to assess skin temperature [21]. Moreover, athletes are presumed to keep their thermal pattern constant in baseline conditions, and thermal asymmetries in the bilateral body areas (e.g., ankle, knee, hamstring or elbow) are linked to factors related to injuries, such as inflammation or secondary trauma, as in normal conditions, the temperature of both sides of the bilateral areas should increase equally [22]. IRT can detect these asymmetries by comparing bilateral body areas, showing potential injury risk due to incorrect work assimilations provoked by factors such as excessive training, poor technique or muscle overload [23].

The relevance of IRT is that it can detect temperature asymmetries (and consequent risks) before other markers such as pain, making this method extraordinarily effective and applicable in preventing injuries [23]. If IRT provides coaches with the ability to detect impairments in bilateral body areas before pain, it can allow them to modify the training load proactively, decreasing injury risk and increasing performance [15]. This advantage is especially remarkable in sports like football, in which high-intensity interval training combined with weekly competition can lead the locomotor system to its anatomical and physiological limit [19], exponentially increasing the risk of minor injuries, overuse injuries, lower-limb injuries and muscle strains [19]. In recent years, authors have explained the efficiency of this technology for injury prevention in medicine [24]; however, IRT has not been widely investigated in athletes such as football players, as only a couple of studies have explored the use of IRT for preventing football injuries [25,26]. Nevertheless, none of the studies performed with thermography have aimed to compare thermography results with the information that external loads provide about players' physical condition in order to adapt their training loads in accordance with the results; therefore, this concept demands research.

In this sense, the aim of the study was to determine if there is a relationship between thermal asymmetries provided by thermography and performance in a repeated sprint ability (RSA) test and in countermovement jumps (CMJs) before and after the RSA test. If thermography can predict these tests´ performance without the need to execute them, it can offer coaches the possibility of adapting players' training loads without the need for them to exert maximal efforts.

2. Materials and Methods

2.1. Experimental Approach to the Problem

This study used repeated measures within participants to determine the relationship between skin temperature and/or thermal asymmetries in the bilateral body areas, CMJ performance and RSA performance. The protocol consisted of a thermography session of the lower limbs (anterior and posterior parts) of each player in basal conditions after 48 h resting, followed by a standardized warm-up that included 5 min of continuous running, 5 min of joint mobility and two sprints of 30 m with a recovery process of 2 min, and a vertical jump test (CMJ) + repeated sprint ability (RSA) test (6 × 30 m/20″) + vertical jump test (CMJ) (See Figure 1 for more clarity).

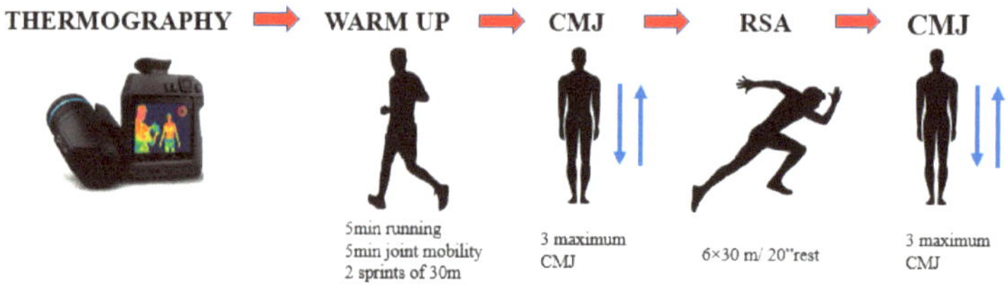

Figure 1. Flowchart of the protocol.

2.2. Participants

The participants were 20 male football players from a professional team of the Smartbank league (the second Spanish football division) (age 28.9 ± 3.9 years; height 178.7 ± 9 cm; body mass 74.8 ± 6.4 kg). Players with recent injuries or pain were excluded, as they could interfere with the results. Each player signed informed consent with an explanation of the study procedure as well as the associated risks.

2.3. Ethical Statement

The study was conducted according to the requirements of the Declaration of Helsinki (2013) and was approved and followed the guidelines stated by the Ethics Committee of the European University of Madrid (CIPI35/2019). The research also received formal approval from the professional football club involved.

2.4. Measures

2.4.1. Thermography

The collection of thermographic data followed the standards proposed by the consensus statement of TISEM on the measurement of human skin temperature [27]. Thermograms were evaluated before the protocol and used as a control variable prior to testing. Thermograms were performed in an air-conditioned room; the temperature was set at 22 °C (±1.5 °C) with about 40–60% of relative humidity, and the skin temperature of the lower limbs at the anterior and posterior parts was recorded. The thermal camera FLIR T420bx (FLIR Systems, Sweden) with a resolution of 320 × 240 pixels was placed 3 m away from the participants and at a perpendicular angle to them. The players were instructed to rest 24 h prior to the thermograms and to avoid behaviours that could interfere with the assessment of thermal images, such as drinking alcohol, smoking or consuming caffeine. During testing, the participants were dressed in underwear and were barefoot, so selected areas of skin were continuously exposed during the exercises and measurements. Following the Thermohuman technology protocol, to facilitate the analysis, the body regions of interest (ROIs) analysed included the thighs, calves and hamstrings, as these are the main muscles of the lower limbs [23]. Computerized image analysis allowed the selection of the measurement area in the thermograms (see Figure 2). The areas were analyzed with the Thermohuman software (PEMA THERMO GROUP, Spain). From the analyzed ROIs, average, minimum, maximal temperature was extracted to calculate thermal asymmetries between bilateral ROIs.

Figure 2. Measurement areas provided by the Thermohuman software (anterior and posterior parts).

The players were gathered into pairs in subgroups (low [<0.2 °C] vs. high thermal asymmetry [≥0.2 °C]) based on the results of the thermography session. Previous literature has considered clinically significant skin temperature asymmetry to be over 0.5 °C, but it was decided to establish the cut-off point at 0.2 °C, because the sample is from a professional club and the members were highly supervised. The physical department of the club made the political decision to start following, taking care of and monitoring players when they have ≥0.2 °C asymmetry. Thus, not many players reached 0.5 °C asymmetry. Therefore, the sample was divided into low and high thermal asymmetry of thighs, hamstrings and calves. The players were grouped in terms of their asymmetry for each muscle group; accordingly, one player could be in the high-asymmetry group for one of the studied muscles and the low-asymmetry group for another muscle.

2.4.2. Vertical Jumps

Players completed 2 CMJ tests: (1) after warming up and (2) after the RSA test, with some minutes to recover from it. To measure the height of the jumps, infrared technology was used (Optojump Next, Microgate, Bolzano, Italy). The participants were already familiar with the movement, and they were instructed to undertake it in the most precise way, keeping their hands on their hips to eliminate the influence of arm movement on the jump performance. Each player performed three jumps before and after the RSA test (with 2 min of recovery between jumps). The average of the three jumps was calculated for the statistical analysis.

2.4.3. Repeated Sprint Ability (RSA)

The RSA test included six sprints of 30 m with 20 s recovery between sprints. Two pairs of photocells (Witty, Microgate, Bolzano, Italy), placed at 0 and 30 m, were used. The following measures were calculated: sprint time (RSAt), best sprint time, average time, total time, percentage difference between the first and the rest of the sprints during the RSA test – %Diff – [((sprint time – first sprint time)/first sprint time) × 100] and percentage difference between the best and the rest of the sprints during the RSA test – %Best – [((sprint time – best time)/best time) × 100]. The previous two sprints performed during the warm-up were used as a control measure to ensure that players performed the RSA test at maximum speed. If the time of the first RSA test sprint was longer (>5%) than the best individual sprint performed before the start of the test, the RSA test was not considered valid, and the player had to repeat the test after 5 min of recovery.

2.5. Statistical Analysis

The data are presented as means ± standard deviations. The normal distribution of the variables was confirmed by histogram charts and the Shapiro–Wilk distribution test. A mixed two-way analysis of variance was performed to analyse the effect of the level of thermal asymmetry on the RSA variables depending on the sprint number. The thermal asymmetry group was used as an independent factor, and the sprint number as a repeated measured factor to control the interaction of the thermal asymmetry group with the sprint progression. Bonferroni post hoc pairwise comparisons were undertaken to compare the RSA variables of the two groups of thermal asymmetries in each of the sprints, including the confidence interval (95%). The effect size (ES) was calculated for all the inference tests using the partial eta squared (ηp2 = ES) value with the following interpretation: small (ES = 0.01–0.059); medium (ES = 0.06–0.14); and large effects (ES > 0.14). The level of significance was set at $p < 0.05$ for all the tests, and all the data were statistically analysed using SPSS V24.0. (IBM Corp. Released 2016. IBM SPSS Statistics for Windows, Version 24.0. Armonk, NY: IBM Corp.)

3. Results

The normal distribution of the variables was confirmed by histogram charts and the Shapiro–Wilk distribution test (the statistic varies between 0.926 and 0.984, $p > 0.005$ in all cases). Figure 3 shows the results for the RSA variables: the sprint time (RSAt), the percentage difference between the first and the rest of the sprints during the RSA test – %Diff – [((sprint time – first sprint time)/first sprint time) × 100] and the percentage difference between the best and the rest of the sprints during the RSA test – %Best – [((sprint time – best time)/best time) × 100]. The results are calculated based on thermal asymmetry clusters (a group with low asymmetry and a group with high asymmetry in each of the three muscle groups). Thigh and calf asymmetry did not show a significant effect on any variable ($p > 0.05$). Moreover, no significant interaction was found between thigh and calf asymmetry and the number of sprints ($p > 0.05$). Conversely, hamstring asymmetry did not show a significant effect in the RSA test and %Diff ($p > 0.05$), but it did in %Best (F = 6.59; $p = 0.018$; ES = 0.25), highlighting fewer differences in high asymmetry with respect to the best sprint. Furthermore, a significant interaction was found between hamstring asymmetry and sprint number 2 in the RSAt (F = 2.42; $p = 0.041$; ES = 0.11) and in %Diff (F = 2.50; $p = 0.049$; ES = 0.11) but not in %Best ($p > 0.05$). Regarding the pair-wise comparison, the hamstring high-asymmetry group showed a higher %Diff in sprint 2 (F = 7.40; $p = 0.013$; IC: −4.20 to −0.55; ES = 0.27), and there was also a higher %Best in the hamstring low-asymmetry group in sprints 2 (F = 5.76; $p = 0.026$; IC: 0.11

to 1.55; ES = 0.22), 3 (F = 15.59; $p = 0.001$; IC: −1.55 to −0.11; ES = 0.44) and 4 (F = 9.36; $p = 0.006$; IC: 0.67 to 2.22; ES = 0.32).

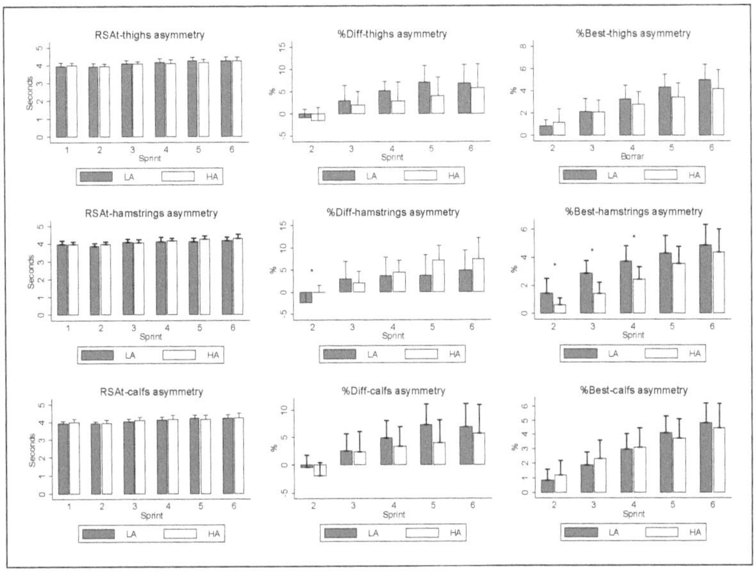

Figure 3. Time differences based on low (LA) and high (HA) thermal asymmetry groups. * Significant differences between groups ($p < 0.05$). (RSAt = sprint time). (%Diff = [((sprint time − first sprint time)/first sprint time) × 100]). (%Best = [((sprint time − best time)/best time) × 100].

Table 1 shows the differences in CMJ variables between all the defined groups. There are no significant differences in any of the variables between the low and the high thermal asymmetry group for thighs and calves. However, for hamstrings, while there are no significant differences between the low and the high thermal asymmetry group in the CMJ Pre-RSA and Diff CMJ tests ($p > 0.05$), the low asymmetry group has significantly higher values for the CMJ Post-RSA (F = 7.55; $p = 0.013$; IC: 1.06 to 7.84; ES = 0.28).

Table 1. Differences in CMJ variables depending on low and high asymmetry.

		Low Asymmetry	High Asymmetry
Thigh asymmetry	CMJ Pre	39.30 ± 3.59	38.78 ± 4.74
	CMJ Post	34.35 ± 4.07	35.13 ± 4.64
	Diff CMJ	−4.96 ± 3.12	−3.65 ± 3.17
Hamstring asymmetry	CMJ Pre	40.80 ± 3.62	37.79 ± 3.95
	CMJ Post	37.23 ± 3.78	32.78 ± 3.59 *
	Diff CMJ	−3.58 ± 3.28	−5.01 ± 3.01
Calf asymmetry	CMJ Pre	38.98 ± 4.41	39.19 ± 3.77
	CMJ Post	34.15 ± 3.97	35.27 ± 4.64
	Diff CMJ	−4.83 ± 2.53	−3.92 ± 3.77

* Significant differences between groups ($p < 0.05$).

4. Discussion

The purpose of this study was to determine if there is a relationship between thermal asymmetries provided by thermography and performance in a repeated sprint ability (RSA) test and in countermovement jumps (CMJs) before and after the RSA test. The results of the RSA test showed a significant interaction between hamstring asymmetry and the number of sprints in the RSAt and in the % Diff. The hamstring high-asymmetry group showed a higher %Diff in sprint 2, which means that there is a significative longer time with respect to sprint 1 (worse performance) than in the low-asymmetry group. On the other hand, the hamstring low-asymmetry group reported a higher %Best in sprints 2, 3 and 4. The cause of this difference in the hamstring %Best may be attributable to the fact that the low-asymmetry group had the best sprint, which was very fast; therefore, the following best ones have a significant difference (worse). On the other hand, no differences were found in the thigh and calf groups for any variable. The reason why the only thermal asymmetry that influenced the performance was that of hamstrings could be their importance in sprinting, as they play a crucial role in generating force in the propulsive part of the sprint [28] and are clearly the most frequently injured muscle during sprinting [4]. The majority of hamstring muscle injuries occur while the athlete is running at maximal or close to maximal speeds [29] Many investigations have measured hamstring activity during sprinting with electromyographic (EMG), and found that the hamstrings are active from the middle of the swing to the final stance Some of these studies have reported that peak activity occurs during terminal swing [30], whereas others have found it to occur during stance [31]. Proof of the effort that these muscles exert in the sprint is that an increase in running speed of 80–100% is linked with an increase in net hamstring muscle force and energy absorption of 1.4 and 1.9 times, respectively [32]. However, this study took data from athletes running on a treadmill, and the mechanical properties of treadmill surfaces are different from those of the surfaces on which the athletes usually train and compete (e.g., artificial turf or an athletic track) [33]. This factor may influence running anatomy, so the results should be interpreted with caution.

Regarding the CMJ results, the low thermal asymmetry hamstring group has significantly higher values in the CMJ post-RSA, while in the calf and thigh groups, there are no significant differences in either CMJ pre- or CMJ post-RSA. Thus, the high thermal asymmetry hamstring group's performance was similar to that of the low thermal asymmetry hamstring group before the RSA, and then its performance decreased in the CMJ post-RSA. This can be indicative of fatigue, as a decreased average CMJ jump height can indicate neuromuscular fatigue [34]. The reason why significant differences are only found in the high-asymmetry hamstring group (and not in the calf and thigh groups) may be that hamstrings are the most implicated and frequently injured group in sprinting [28], while calves and thighs seem to play a less crucial role in sprinting; therefore, asymmetry takes longer to turn into fatigue when performing an RSA test, as they are less implicated in this activity.

Although the results of this study seem to provide evidence in terms of using and interpreting thermography to assess RSA and CMJ performance in elite football players, the results must be interpreted with caution, as certain limitations should be considered. First, thermography was carried out in just one session; applying thermography in a higher number of sessions would allow us to corroborate the results of the study. Second, a low asymmetry cut point was established, and settling a higher asymmetry cut point would probably show stronger correlations. Finally, the sample was very small, so future research should be carried out to look for evidence and to check whether these results apply to different genders, ages and competitive levels.

From a practical point of view, if football coaches can access the thermographs of their players, they should focus especially on hamstrings (rather than calves and thighs) and modify the training if necessary. These muscles appear to be the ones with more influence on the test results, as they are the only ones for which thermal asymmetry of over 0.2 °C had an effect.

5. Conclusions

Thermography can provide information about performance in CMJ and RSA tests for hamstring asymmetries over 0.2 °C; meanwhile, calf and thigh asymmetries over 0.2 °C do not seem to have a relationship with performance in these tests. In short, thermography results have some correlation with CMJ and RSA performance, but stablishing higher cut off points would probably allow us to find stronger correlations for hamstring asymmetries, and maybe some correlation for calf and thigh asymmetries which would be helpful for coaches and athletes.

Author Contributions: Conceptualization, C.M. and J.L.F.; methodology, C.M. and J.L.F.; software, J.G.-U. and A.H.-M.; validation, J.S.-S. and L.G.; formal analysis, C.M. and J.L.F..; investigation, C.M., J.L.F. and A.H.-M.; resources, J.S.-S. and L.G.; data curation, J.G.-U.; writing—original draft preparation, C.M. and J.L.F..; writing—review and editing. J.L.F. and A.H.-M.; visualization, J.L.F. and A.H.-M.; supervision, J.S.-S. and L.G.; project administration, L.G. All authors have read and agreed to the published version of the manuscript.

Funding: This research received no external funding.

Institutional Review Board Statement: The research also received formal approval from the professional football club involved. The study was conducted according to the requirements of the Declaration of Helsinki (2013) and was approved and followed the guidelines stated by the Ethics Committee of the European University of Madrid (CIPI35/2019). The research also received formal approval from the professional football club involved.

Informed Consent Statement: Not applicable.

Data Availability Statement: The data presented in this study is available on request from the corresponding author.

Acknowledgments: C.M. acknowledges the Spanish Ministry of Science, Innovation and Universities for funding the development of his Ph.D. (Grant Number: FPU20/00243). A.H.-M. acknowledges the Spanish Ministry of Science, Innovation and Universities for funding the development of his Ph.D. (Grant Number: FPU18/03222).

Conflicts of Interest: The authors declare no conflict of interest.

References

1. Rebelo, A.; Brito, J.; Maia, J.; Coelho-e-Silva, M.; Figueiredo, A.; Bangsbo, J.; Malina, R.M.; Seabra, A. Anthropometric characteristics, physical fitness and technical performance of under-19 soccer players by competitive level and field position. *Int. J. Sport. Med.* **2013**, *34*, 312–317. [CrossRef] [PubMed]
2. Iaia, F.M.; Ermanno, R.; Bangsbo, J. High-intensity training in football. *Int. J. Sport. Physiol. Perform.* **2009**, *4*, 291–306. [CrossRef] [PubMed]
3. López-Valenciano, A.; Ruiz-Pérez, I.; Garcia-Gómez, A.; Vera-Garcia, F.J.; Croix, M.D.S.; Myer, G.D.; Ayala, F. Epidemiology of injuries in professional football: A systematic review and meta-analysis. *Br. J. Sport. Med.* **2020**, *54*, 711–718. [CrossRef] [PubMed]
4. Ekstrand, J.; Hägglund, M.; Waldén, M. Epidemiology of muscle injuries in professional football (soccer). *Am. J. Sport. Med.* **2011**, *39*, 1226–1232. [CrossRef]
5. Halson, S.L. Monitoring training load to understand fatigue in athletes. *Sport. Med.* **2014**, *44*, 139–147. [CrossRef]
6. Ekstrand, J. Epidemiology of football injuries. *Sci. Sport.* **2008**, *23*, 73–77. [CrossRef]
7. Gabbett, T.J.; Jenkins, D.G. Relationship between training load and injury in professional rugby league players. *J. Sci. Med. Sport* **2011**, *14*, 204–209. [CrossRef]
8. Alexiou, H.; Coutts, A.J. A comparison of methods used for quantifying internal training load in women soccer players. *Int. J. Sport. Physiol. Perform.* **2008**, 3, 320–330. [CrossRef]
9. Buchheit, M.; Racinais, S.; Bilsborough, J.; Bourdon, P.; Voss, S.; Hocking, J.; Cordy, J.; Mendez-Villanueva, A.; Coutts, A. Monitoring fitness, fatigue and running performance during a pre-season training camp in elite football players. *J. Sci. Med. Sport* **2013**, *16*, 550–555. [CrossRef]
10. Coyne, J.O.C.; Gregory Haff, G.; Coutts, A.J.; Newton, R.U.; Nimphius, S. The Current State of Subjective Training Load Monitoring—A Practical Perspective and Call to Action. *Sport. Med.—Open* **2018**, *4*, 58. [CrossRef]
11. Chuckravanen, D.; Bulut, S.; Kürklü, G.; Yapali, G. Review of exercise-induced physiological control models to explain the development of fatigue to improve sports performance and future trend. *Sci. Sport.* **2019**, *34*, 131–140. [CrossRef]
12. Turner, A.N.; Stewart, P.F. Repeat sprint ability. *Strength Cond. J.* **2013**, *35*, 37–41. [CrossRef]
13. Brownstein, C.G.; Dent, J.P.; Parker, P.; Hicks, K.M.; Howatson, G.; Goodall, S.; Thomas, K. Etiology and recovery of neuromuscular fatigue following competitive soccer match-play. *Front. Physiol.* **2017**, *8*, 831. [CrossRef]
14. Morcillo, J.A.; Jiménez-Reyes, P.; Cuadrado-Peñafiel, V.; Lozano, E.; Ortega-Becerra, M.; Párraga, J. Relationships between repeated sprint ability, mechanical parameters, and blood metabolites in professional soccer players. *J. Strength Cond. Res.* **2015**, *29*, 1673–1682. [CrossRef]

15. Watkins, C.M.; Barillas, S.R.; Wong, M.A.; Archer, D.C.; Dobbs, I.J.; Lockie, R.G.; Coburn, J.W.; Tran, T.T.; Brown, L.E. Determination of vertical jump as a measure of neuromuscular readiness and fatigue. *J. Strength Cond. Res.* **2017**, *31*, 3305–3310. [CrossRef]
16. Balsalobre-Fernández, C.; Tejero-González, C.M.; del Campo-Vecino, J. Relationships between training load, salivary cortisol responses and performance during season training in middle and long distance runners. *PloS ONE* **2014**, *9*, e106066. [CrossRef]
17. Morin, J.-B.; Dupuy, J.; Samozino, P. Performance and fatigue during repeated sprints: What is the appropriate sprint dose? *J. Strength Cond. Res.* **2011**, *25*, 1918–1924. [CrossRef]
18. Gathercole, R.; Sporer, B.; Stellingwerff, T.; Sleivert, G. Alternative countermovement-jump analysis to quantify acute neuromuscular fatigue. *Int. J. Sport. Physiol. Perform.* **2015**, *10*, 84–92. [CrossRef]
19. Hildebrandt, C.; Zeilberger, K.; Ring, E.F.J.; Raschner, C. The application of medical infrared thermography in sports medicine. *Int. Perspect. Top. Sport. Med. Sport. Inj.* **2012**, *534*, 257–274.
20. Fernández Cuevas, I. Effect of Endurance, Speed and Strength Training on Skin Temperature Measured by Infrared Thermography = Efecto del Entrenamiento de Resistencia, Velocidad y Fuerza en la Temperatura de la Piel a Través de la Termografía Infrarroja. Ph.D. Thesis, Facultad de Ciencias de la Actividad Física y del Deporte, Madrid, Spain, 2012.
21. Alfieri, F.M.; Battistella, L.R. Body temperature of healthy men evaluated by thermography: A study of reproducibility. *Technol. Health Care* **2018**, *26*, 559–564. [CrossRef]
22. Fernández-Cuevas, I.; Marins, J.C.B.; Lastras, J.A.; Carmona, P.M.G.; Cano, S.P.; García-Concepción, M.Á.; Sillero-Quintana, M. Classification of factors influencing the use of infrared thermography in humans: A review. *Infrared Phys. Technol.* **2015**, *71*, 28–55. [CrossRef]
23. Fernández-Cuevas, I.; Lastras, J.A.; Galindo, V.E.; Carmona, P.G. Infrared thermography for the detection of injury in sports medicine. In *Application of Infrared Thermography in Sports Science*; Springer: Berlin, Germany, 2017; pp. 81–109.
24. Barcelos, E.Z.; Caminhas, W.M.; Ribeiro, E.; Pimenta, E.M.; Palhares, R.M. A combined method for segmentation and registration for an advanced and progressive evaluation of thermal images. *Sensors* **2014**, *14*, 21950–21967. [CrossRef] [PubMed]
25. Gómez-Carmona, P.; Fernández-Cuevas, I.; Sillero-Quintana, M.; Arnaiz-Lastras, J.; Navandar, A. Infrared thermography protocol on reducing the incidence of soccer injuries. *J. Sport Rehabil.* **2020**, *29*, 1222–1227. [CrossRef] [PubMed]
26. Côrte, A.C.; Pedrinelli, A.; Marttos, A.; Souza, I.F.G.; Grava, J.; Hernandez, A.J. Infrared thermography study as a complementary method of screening and prevention of muscle injuries: Pilot study. *BMJ Open Sport Exerc. Med.* **2019**, *5*, e000431. [CrossRef] [PubMed]
27. Moreira, D.G.; Costello, J.T.; Brito, C.J.; Adamczyk, J.G.; Ammer, K.; Bach, A.J.; Costa, C.M.; Eglin, C.; Fernandes, A.A.; Fernández-Cuevas, I. Thermographic imaging in sports and exercise medicine: A Delphi study and consensus statement on the measurement of human skin temperature. *J. Therm. Biol.* **2017**, *69*, 155–162. [CrossRef]
28. Howard, R.M.; Conway, R.; Harrison, A.J. Muscle activity in sprinting: A review. *Sport. Biomech.* **2018**, *17*, 1–17. [CrossRef]
29. Schache, A.G.; Dorn, T.W.; Blanch, P.D.; Brown, N.A.; Pandy, M.G. Mechanics of the human hamstring muscles during sprinting. *Med. Sci. Sport. Exerc.* **2012**, *44*, 647–658. [CrossRef]
30. Higashihara, A.; Ono, T.; Kubota, J.; Okuwaki, T.; Fukubayashi, T. Functional differences in the activity of the hamstring muscles with increasing running speed. *J. Sport. Sci.* **2010**, *28*, 1085–1092. [CrossRef]
31. Kyröläinen, H.; Avela, J.; Komi, P.V. Changes in muscle activity with increasing running speed. *J. Sport. Sci.* **2005**, *23*, 1101–1109. [CrossRef]
32. Chumanov, E.S.; Heiderscheit, B.C.; Thelen, D.G. The effect of speed and influence of individual muscles on hamstring mechanics during the swing phase of sprinting. *J. Biomech.* **2007**, *40*, 3555–3562. [CrossRef]
33. Colino, E.; Felipe, J.L.; Van Hooren, B.; Gallardo, L.; Meijer, K.; Lucia, A.; Lopez-Fernandez, J.; Garcia-Unanue, J. Mechanical properties of treadmill surfaces compared to other overground sport surfaces. *Sensors* **2020**, *20*, 3822. [CrossRef]
34. Claudino, J.G.; Cronin, J.; Mezêncio, B.; McMaster, D.T.; McGuigan, M.; Tricoli, V.; Amadio, A.C.; Serrão, J.C. The countermovement jump to monitor neuromuscular status: A meta-analysis. *J. Sci. Med. Sport* **2017**, *20*, 397–402. [CrossRef]

sensors

MDPI

Article

From Optical Tracking to Tactical Performance via Voronoi Diagrams: Team Formation and Players' Roles Constrain Interpersonal Linkages in High-Level Football

Nelson Caldeira [1,*], Rui J. Lopes [2,3], Dinis Fernandes [3] and Duarte Araujo [1]

1 CIPER, Faculty of Human Kinetics, University of Lisbon, 1499-002 Cruz Quebrada, Portugal
2 Department of Information Science and Technology (ISTA), Iscte—Instituto Universitário de Lisboa, 1649-026 Lisbon, Portugal
3 Instituto de Telecomunicações, 1049-001 Lisbon, Portugal
* Correspondence: coach.caldeira@gmail.com

Abstract: Football performance behaviour relies on the individual and collective perceptual attunement to the opportunities for action (affordances) available in a given competitive environment. Such perception–action coupling is constrained by players' spatial dominance. Aiming to understand the influence of team formation and players' roles in their dynamic interaction (interpersonal linkages), Voronoi diagrams were used to assess the differences in players' spatial dominance resulting from their interactions according to ball-possession status in high-performance football. Notational (i.e., team formation, players' role, and ball-possession status) and positional data (from optical sensors) from ten matches of the men's French main football league were analysed. Voronoi diagrams were computed from players' positional data for both teams. Probability density functions of the players' Voronoi cell areas were then computed and compared, using the Kolmogorov–Smirnov test, for the different variables (i.e., team formation, player role, and ball-possession status) and their classes. For these variables, the players' Voronoi cell areas presented statistical differences, which were sensitive to team formation classes (i.e., defenders, midfielders, and forwards) and relative pitch location (interior or exterior in the effective play space). Differences were also found between players with similar roles when in different team formations. Our results showed that team formation and players' roles constrain their interpersonal linkages, resulting in different spatial dominance patterns. Using positional data captured by optical sensors, Voronoi diagrams can be computed into compound variables, which are meaningful for understanding the match and thus offer information to the design representative training tasks.

Keywords: affordances; spatial dominance patterns; performance; team synergies; Voronoi cells

Citation: Caldeira, N.; Lopes, R.J.; Fernandes, D.; Araujo, D. From Optical Tracking to Tactical Performance via Voronoi Diagrams: Team Formation and Players' Roles Constrain Interpersonal Linkages in High-Level Football. *Sensors* **2023**, *23*, 273. https://doi.org/10.3390/s23010273

Academic Editors: Bruno Gonçalves, Bruno Emanuel Nogueira Figueira and Diogo Alexandre Martins Coutinho

Received: 12 November 2022
Revised: 21 December 2022
Accepted: 23 December 2022
Published: 27 December 2022

1. Introduction

In recent years, the technological progress around spatial location systems and positional data has had a growing impact on our societies and in all investigation fields [1,2], including sport sciences [3] and high-performance football [4].

This increase in the volume of data [5] can better inform coaches about the performance of their teams, including tactical behaviour [6]. Importantly, teams' performance is based on the coordinated decisions of their players [7], which form team synergies (spatial-temporal patterns of coordination) guided by shared affordances [8].

Affordances are properties of the environment that relate to the individual characteristics, implemented by specific perception–action cycles, i.e., "action specific relations that exist between the skills/capacities of an individual performer and the action relevant properties of a [perceived environmental] task" ([8] p. 4). Training develops football players to become attuned to affordances of the match, namely, those constrained by match phases such as ball possession status [9,10]. Such *attunement* is better developed if coaches pursue the *representativeness* of their training exercises [11] through the manipulations of relevant task constraints [12].

When training for a match, coaches constrain the emergence of the affordances perceived by players and, consequently, their interactions with teammates and opponents [13]. For this purpose, there are evidence-based match-space criteria for training design. For example, coaches can define the space of their training exercises from generic benchmarks such as the *Game Intensity Index* (GII) [14]. The GII establishes a parallelism of the training surface in terms of square meters per player with that of competition. However, this is a very broad reference, which, in high-competition football, is equivalent to an area of 325 m^2 per player (68 m × 105 m/22 *players*). It is no surprise that studies with *small-sided and conditioned games* (SSCG) suggest the use of *relative space per player* (RSP). The RSP corresponds to an area per player that derives from the smallest rectangle where all field players fit [15]. Similarly, Silva and colleagues [16] divided the *effective play space* (EPS), which is the polygon of the smallest convex hull, by the number of players. Both RSP proposals have the merit of measuring what occurs in game spaces in training and competition. However, they do not consider the space outside the EPS and, consequently, the impact of team formation on players' and teams' metrics.

During a football game, players do not move randomly throughout the space [17]. Players' movements and team coordination [18] are constrained by strategy [7], including the game system or *team formation* [19]. These formations constrain the spatial organisation of players in a team [20] and, thus, how they can form synergies [18]. Team formations are especially relevant to understand *interpersonal linkages* as "the specific contribution of each element to a group task" ([8] p. 8). Team formations are typically represented via a set of three or four numbers that indicate the number of players in each line (or sector) and express how the team is organised on the pitch. For example, "3-5-2" expresses that the team formation is composed of three defenders, five midfielders, and two attackers [20]. Moreover, each player in his/her sector has a specific spatial *role* [21,22] or playing position [21], which is tagged with a specific designation. Usually, it describes the player's main role, considering both the sector to which they belong (e.g., defenders—all outfield players that are more implicated in defensive tasks) and information about their corridor and side (e.g., right lateral defender or left centre midfielder). Currently, in high-performance football, 3-5-2, 3-4-1-2, and 4-2-3-1 are among the most commonly used tactical team formations [6,23].

Team formation affects performance by, for example, influencing key performance indicators (KPIs) such as the *Effective Play Space* (EPS) or *Team Separateness* [22]. Although the clear influence of team formations and players' role in individual and team performance, its relevance for understanding the synergies that emerge from players interaction in competition are still unclear.

Aiming to bridge this gap, we argue that if the players' roles within a team formation influence team synergies, then it will be possible to identify their specific contributions. Nowadays, we can compute positional data obtained by different types of sensors (e.g., optical tracking, GPS, or RFID) and calculate team spatial–temporal patterns such as *spatial heatmaps* [24], *major ranges* [25], or *Voronoi diagrams* (VD). VD in particular assess players' *space dominance* [26], as well as, at the team synergetic level, their interpersonal linkages ([8] p. 8), (e.g., maintain ball possession).

This paper aims to understand the influence of team formation and players' roles in the players' spatial dominance resulting from the dynamic interaction (interpersonal linkages) of both teams. Therefore, we expect the following:

- Interpersonal linkages among players are expressed by their spatial interactions and are constrained by team formation and players' roles.
- Players' spatial dominance could be operationalised by Voronoi diagrams (and related spatial statistics), which could capture differences according to team formations, players' roles, and ball-possession status.

2. Materials and Methods

2.1. Data Sources

The data used in this paper were provided by STATS© and obtained through their systems of semi-automatic tracking [27] in ten *Ligue 1* matches (France) of the 2019–2020 season. Data were composed of players' positional data (longitudinal and lateral coordinates) sampled at 10 Hz, and notational data describing match events (representing players' contacts with the ball) and possession episode (PE) information (initial and final instants, team with ball possession).

2.2. Data Processing

The raw data were processed before analysis using the following procedures:

- For each match, the determination of team formations was performed in two steps:
    1. Using the STATS Edge Analysis application:
        a. The match time was divided into six periods of 15 minutes, as suggested by Duarte and colleagues [28]; each period was subdivided in case there was a substitution.
        b. The average longitudinal and lateral position of each player was computed throughout each time period.
    2. From these results and following the suggestion of Carling [29] and Bradley and colleagues [30], a panel of experts identified both team formations during each of the match intervals. The panel was composed of five coaches with at least ten years of professional experience at the highest level and holding an UEFA PRO certification.
- Team formation, players' roles, and ball-possession status were considered crucial to data analysis in this paper; consequently, matches and periods within the matches were grouped and selected according to the following criteria:
    a. For each match, there was an analysed team and an opponent team. For all matches and time periods, the opponent team was always the same and organised under the same team formation (3-5-2). The analysed team was always a different one, forming two groups of five matches. In one group, the analysed team played mostly in a 4-2-3-1 team formation, and in the other group mostly with a 3-4-1-2 team formations.
    b. Within each match, only periods in which teams maintained their team formation (4-2-3-1 or 3-4-1-2 for the analysed team and 3-5-2 for the opponent team) were used. All other periods, either where teams played with different formations or where they were not complete (e.g., after a red card), were discarded.
- Match periods were further filtered so that only open plays were considered; i.e., set plays and time gaps without play were discarded. Each open play was subdivided into ball-possession episodes (PEs). Each PE starts at the instant when a team recovers the ball and ends when that team loses control of the ball. According to STATS© reference manual [31], at least two consecutive events were necessary to form a PE. Each PE was classified, given the analysed team's ball possession status, as *in possession* or *out of possession*. The 4-2-3-1 formation comprised 999 possession episodes (499 in possession, 500 out of possession), whilst the 4-2-3-1 formation comprised 1199 possession episodes (601 in possession, 598 out of possession).
- The role of each player of the analysed team was classified according to his spatial average position in the respective team formation. Table 1, adapted from Riezebos [21], identifies these roles for the two team formations: 4-2-3-1 and 3-4-1-2.
- The average value of the Voronoi cell area (VA) during the PE was computed for each player of the analysed team. VAs are computed at each time frame, using the procedures described by Kim [32], considering all the players of both teams.

Figure 1 illustrates the Voronoi diagrams (VDs) obtained from players' roles with different analysed team formations. Although VDs are computed at each time frame, in Figure 1, each player is represented at the average position along the longitudinal and lateral axes for the five matches considered, and their Voronoi area is circumscribed by dashed lines. Players from the analysed teams (in blue) are indicated using their role tag.

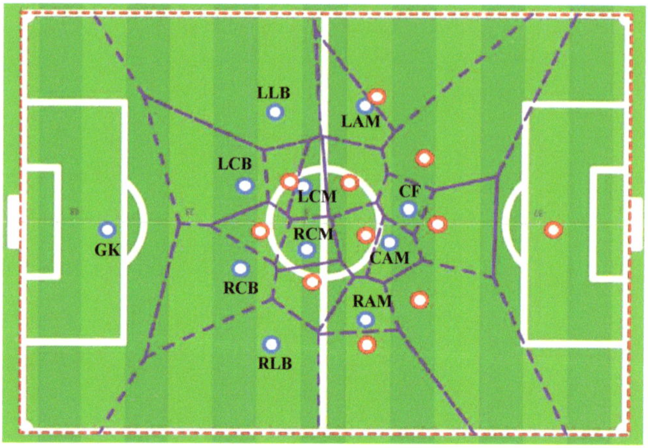

(**a**) 4−2−3−1 Team Formation

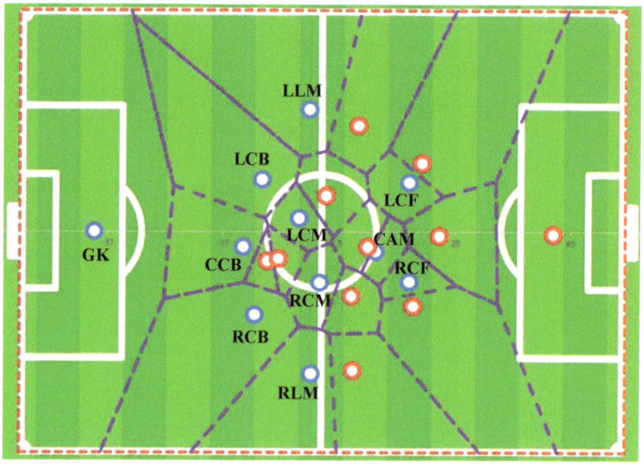

(**b**) 3−4−1−2 Team Formation

Figure 1. Voronoi diagrams for analysed teams (blue) in two team formations ((**a**) 4-2-3-1 and (**b**) 3-4-1-2) and the opposing team (red).

The VDs in Figure 1 exemplify the influence of the opponent team in the VA of the analysed team players, thus capturing the interaction of all players on the pitch and not only with his teammates [33]. This interaction is crucial for the relative position of a given player in the effective play space (EPS). For example, the *Right Centre Forward* (RCF) and the *Left Centre Forward* (LCF) in the 3-4-1-2 team formation, despite being the most forward elements of their team, occupy interior areas in the game space due to the interaction with their opponents.

Table 1. Player's role in 4-2-3-1 and 3-4-1-2 team formations.

4-2-3-1		3-4-1-2	
Tag	*Description*	*Tag*	*Description*
GK	Goalkeeper	GK	Goalkeeper
LLB	Left Lateral Back	CCB	Centre Central Back
LCB	Left Central Back	LCB	Left Central Back
RCB	Right Central Back	RCB	Right Central Back
RLB	Right Lateral Back	LLM	Left Lateral Midfielder
LCM	Left Centre Midfielder	LCM	Left Centre Midfielder
RCM	Right Centre Midfielder	RCM	Right Centre Midfielder
LAM	Left Attacking Midfielder	RLM	Right Lateral Midfielder
CAM	Centre Attacking Midfielder	CAM	Centre Attacking Midfielder
RAM	Right Attacking Midfielder	LCF	Left Centre Forward
CF	Centre Forward	RCF	Right Centre Forward

2.3. Statistical Analysis Methods

Statistical analysis was performed by computing and comparing the probability density functions of the players' mean Voronoi areas (VAs) during each possession episode (PE). Probability density functions are represented as violin plots using kernel smoothing and compared using the Kolmogorov–Smirnov statistic. For the Kolmogorov–Smirnov test, "H0: same distribution" is used as the null hypothesis, with a significance level set at 0.05 (i.e., the alternative hypothesis is assumed if $p < 0.05$).

3. Results

3.1. Comparing Players' Voronoi Areas (VA) within the Same Team Formation (TF)

The results of comparing players' *Voronoi Areas* (VA), according to their role and ball possession status (in possession and out of possession) within the 4-2-3-1 *team formation* (TF), are presented in the violin plots (a) and heatmaps (b and c) of Figure 2. The values in Figure 2 correspond to the Kolmogorov–Smirnov statistic values quantifying the differences between VAs and their statistical significance.

In Figure 2, the value indicated for each role i was computed as $V_{KS}(i) = -log(KS(P_i, Q_i))$ where KS is the Kolmogorov–Smirnov statistic and P_i,Q_i are the VA probability density functions for a player with role i when the analysed team is *in* and *out* of ball possession, respectively. Differences between ball possession status were statistically significant ($p < 0.05$) except for roles highlighted in bold.

The differences between VAs for all possible role pairs are represented in the heatmaps of Figure 2. The value indicated in cell i, j was computed as $V_{KS}(i, j) = -log(KS(P_i, Q_j))$ where KS is the Kolmogorov–Smirnov statistic and P_i, Q_j are, respectively, the VA probability density functions of players with role i and j. Differences between role pairs were statistically significant ($p < 0.05$) except for pairs highlighted in bold.

Figure 2 clearly expose the differences in the distribution of players' VA, according to their role and ball possession. Apart from the Goalkeeper's (GK) specific case, the violin plots also differentiate players' roles according to their sector (back vs. forward) and to their relative location (interior vs. exterior) in the *Effective Play Space* (EPS). Despite the general trend to find significant differences in the VA of players with different roles, ball-possession status also has an impact on VA similarity, mainly in the cases where differences were non-significant.

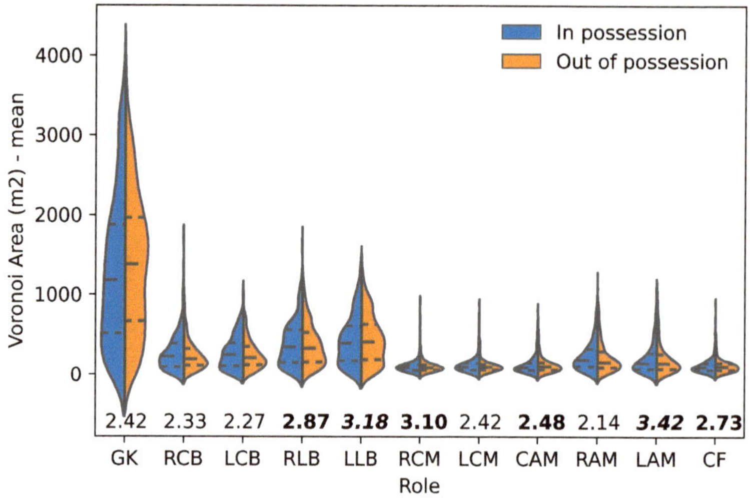

(**a**) Violin plots (in/out of possession).

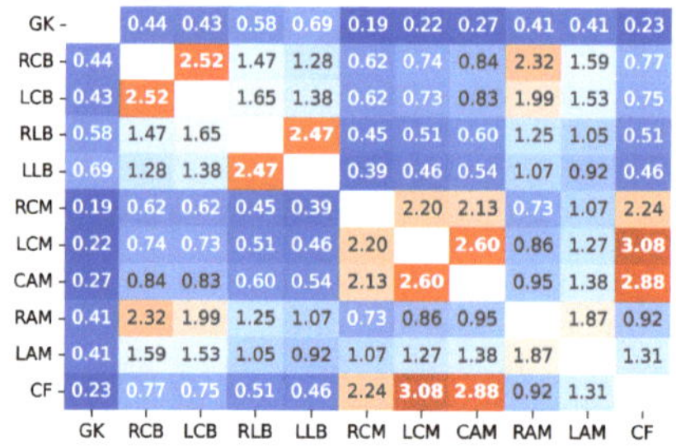

(**b**) KS heatmap (in possession).

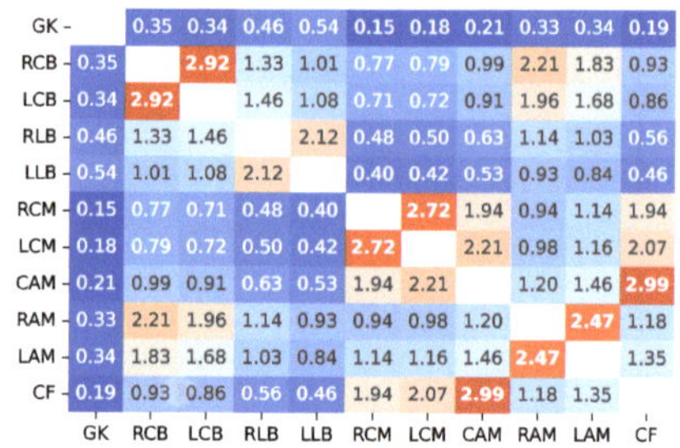

(**c**) KS heatmap (out of possession).

Figure 2. Players' Voronoi area probability density function in the 4-2-3-1 team formation. *Note:* Violin plots (**a**) and heatmaps (**b**,**c**) comparing players' Voronoi area probability density function (less similar in blue, more similar in red). (The differences that are statistically not relevant are highlighted in **bold**).

When the analysed teams were in possession of the ball, the non-significant differences were observed between players of the same sector, namely, between central backs (RCB and LCB), lateral defenders (RLB and LLB), and midfielders (CAM and LCM), and also between the Centre Forward (CF) and two midfield interior players (CAM and LCM). On the other hand, when the analysed teams were out of ball possession, non-significant differences remained between RCB and LCB and between CAM and CF. Except for the new non-significant differences between the defensive midfielders (RCM and LCM) and between wingers (RAM and LAM), all the others were now statistically significant.

The same process was applied to the 3-4-1-2 team formation (TF). Players' VA distribution is presented in the violin plots and their Kolmogorov-Smirnov statistics heatmaps in Figure 3.

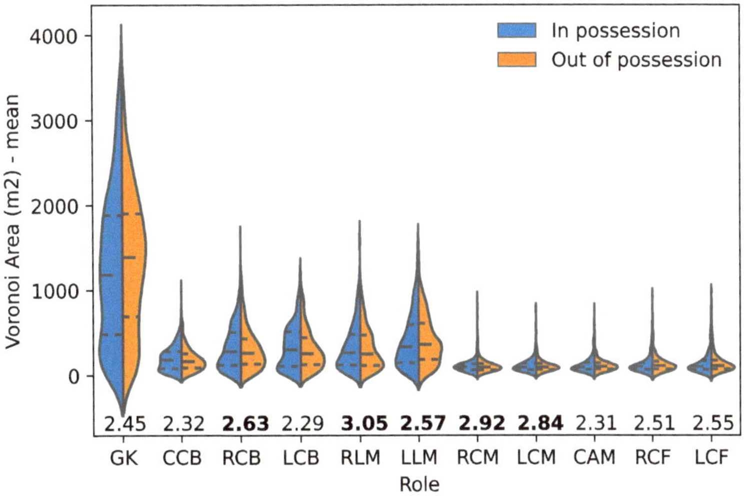

(**a**) Violin plots (in/out of possession).

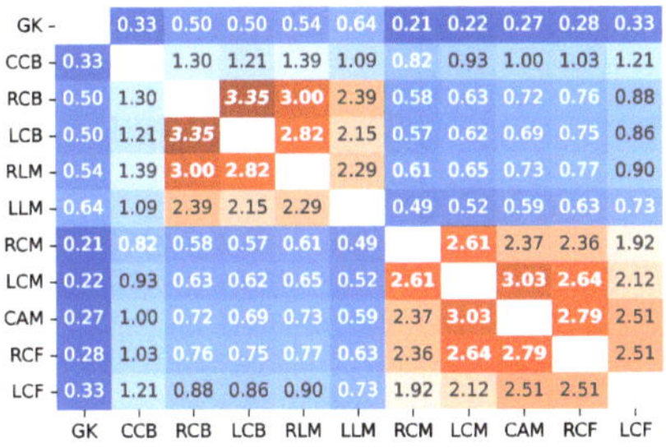

(**b**) KS heatmap (in possession).

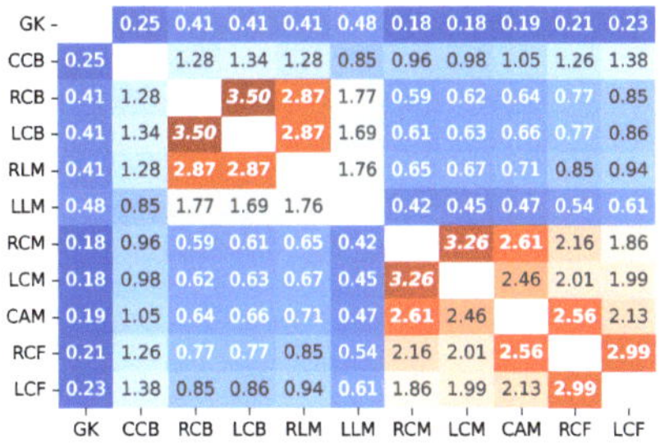

(**c**) KS heatmap (out of possession).

Figure 3. Players' Voronoi area probability density function in the 3-4-1-2 team formation. (*Note:* Violin plots (**a**) and heatmaps (**b**,**c**) comparing players' Voronoi area probability density function (less similar in blue, more similar in red). The differences that are statistically not relevant are highlighted in **bold**).

Similarly, to the 4-2-3-1 team formation, smaller VAs were found for players who usually play in the interior regions of the EPS (RCM, LCM, CAM, RCF, and LCF). In addition, the third central back (CCB) seems to have even smaller areas than the players of the first defensive line (RCB and LCB).

Once more, the ball-possession status significantly influenced only some of the roles (GK, CCB, LCB, CAM, RCF, and LCF). All the other roles presented non-significant differences (highlighted in bold).

In this TF, VA distribution continues to generally allow the differentiation between players' roles. However, the number of non-significant differences increases, showing a higher similarity between the VA of different roles, e.g., in the defensive line (CCB being the exception). The VA differences between role pairs were statistically significant ($p < 0.05$), with the following exceptions (highlighted in bold in Figure 3):

1. In possession: RCB–LCB; RCB–RLM; LCB–RLM; RCM –LCM; LCM–CAM; LCM–RCF; CAM–RCF;
2. Out of possession: RCB–LCB; RCB–RLM; LCB–RLM; RCM–LCM; RCM–CAM; RCF–CAM; RCF–LCF.

3.2. Comparing Players' Voronoi Areas (VA) between Different Team Formations (TF)

Finally, the distribution of VA was compared according to players' roles between team formations (Figure 4).

Both comparisons reveal a natural similarity between players in the same sector. However, the degree of similarity is not equal across sectors and shows important differences between defenders, midfielders, and attackers. In fact, in the defensive sector, all central backs showed significant differences between the TFs. In the lateral backs, the only non-significant difference was found between the LB and the RM. In the lateral backs, the only non-significant difference was found between the LB and the RM.

Concerning the midfielders, distinct analyses should be made for the interior and exterior players. In fact, several non-significant differences were found between the two TFs regarding the interior roles, showing that, for this sub-group, the choice between the 4-2-3-1 and 3-4-1-2 formation did not have a big influence on players' VA, regardless of the team's ball-possession status.

However, for the wingers (RAM and LAM) of the 4-2-3-1 formation, a clear different spatial pattern was found in these players' VA distribution, which does not resemble any other role in the 3-4-1-2 TF (with the exception of the CCB).

Finally, for the attackers, several non-significant differences were found in the comparisons of the central forwards' VA. These are found between players of the attacking sector from the two TFs and between players of the attacking and midfield sectors. This was especially evident in the comparisons with the attacking midfielders (CAM).

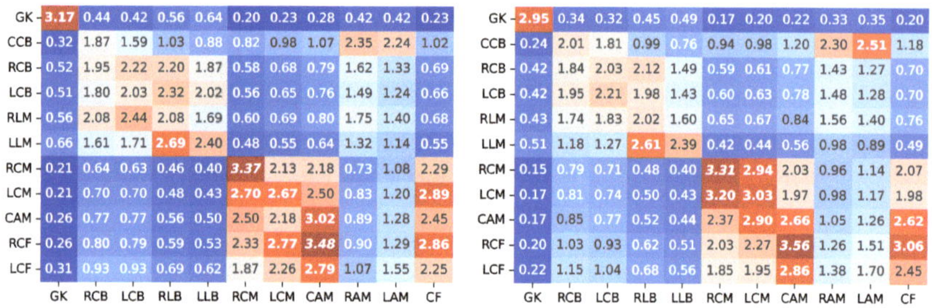

(**a**) KS heatmap (in possession). (**b**) KS heatmap (out of possession).

Figure 4. Comparison of the VA of players' roles in both team formations.

4. Discussion

The aim of this study was to understand the influence of *team formation* (TF) and players' roles in their dynamic interaction (interpersonal linkages). For this purpose, *Voronoi diagrams* (VD) were used to assess the differences in players' spatial dominance resulting from their interactions according to ball-possession status in high-performance football teams.

The observed results support some important reflections on the division of space, according to TF, players' roles, and the ball-possession status. When analysing the spatial patterns within the same TF, the differences between players' VA were generally found to be statistically significant according to their roles.

The results from Fonseca and colleagues [9], showing the influence of ball-possession status on players' *Voronoi cell areas* (VA), do not universally apply but are dependent on *players' roles*. With few some exceptions, these differences demonstrate the existence of different affordances according to players' roles, especially between *sectors* and according to the relative location (*interior* or *exterior*) of each player in the *Effective Play Space* (EPS). The VA of each player role varies according to these two general criteria, influencing players' interpersonal linkages in the establishment of collective synergies.

Resulting from the teams' interactions and due to the nature of the VA metric, players who usually act in the interior of the EPS (midfielders and centre forwards) had smaller areas; players who play in the periphery of the EPS had larger VAs; and the VAs of the wingers or attacking midfielders (RAM and LAM) were intermediate (possibly because they alternate between interior and exterior spaces in the EPS). Additionally, more defensive players (who occupy positions closer to their own goal) generally deal with wider VAs, while the most offensive players usually deal with smaller VAs.

Although some role pairs show a certain degree of symmetry (RCB–LCB or RCM–LCM), this is not found globally. This is the case for the attacking midfielders (AMR–AML) in the 4-2-3-1 TF or the players of the lateral corridors (RM–LM) in the 3-4-1-2 TF. In comparing the two TFs analysed, it is important to note how VA patterns significantly change across the players of the defensive sector, exposing differences in the spatial affordances when a team plays with a first defensive line with three or four players. However, other roles were very similar in both team formations. Apart from the expected case of the GK, this similarity was especially true in the interior (centre) midfield positions. Our results also underpin some practical clues that we find relevant to the football coaches' work. First is the need to consider TF and players' roles as important constraints during the design of training exercises [19]. This implies the need to manipulate and measure the space per player role that actually occurs within these drills in reference to the competitive patterns [18].

Additionally, the fact that no significant differences were found between some players' roles (e.g., between the two central backs or the two more defensive midfielders—a fact found in both TFs and independent of ball possession status) may indicate that players can eventually switch more easily between these roles. This is particularly relevant in situations out of the game plan, e.g., when replacing an injured player.

However, as most role-pair comparisons presented significant differences, coaches need to be aware of the difficulty for players to adapt to the spatial affordances associated with different roles. Even within the same sector, switching sides may imply different spatial affordances due to the non-symmetry detected in some roles. The difference between players' role patterns that were expected to be similar between the two TFs highlights the need for coaches to dedicate enough training time to attune players, individually and collectively, to the spatial affordances that emerge from the strategical option for a given TF [34]. Coaches should be aware that a sudden change in TF may cause more difficulties in adapting to their players than to their opponents. Moreover, differences in VA spatial patterns, according to the TF and players' roles, may also imply the need to properly consider them in the long-term training processes of youth footballers. For example, by introducing a certain degree of variability in the role, coaches can avoid a possible early specialisation [35].

Voronoi diagrams can thus be considered a useful tool to study teams in competitions (match analysis) and as an auxiliary metric to the design of representative training exercises. After characterising the VA of each player role in a given team organisation (TF) during competition, the next step is to use the same tools in training exercises. By measuring players' VA in each exercise, it will be possible to compare the data obtained in the context of training with the values of the respective team in the context of competition. This can constitute a possible way to quantify, in spatio-temporal terms, their representativeness degree, with more detail than with *Game Intensity Index* [14] or the *Relative space per player* [15,16]. In particular, and contrary to the relatively simplistic idea proposed in [36] that 320 m^2 per player would be more representative to design Small-Sided and Conditioned Games (SSCG), VD can help coaches to more effectively manipulate training surfaces. In fact, the adoption of VD to assess players' spatial patterns can help in the definition of more suitable dimensions for each training exercise, adjusting them to the global TFs and players' roles. The use of VD-based tools can contribute to achieving a higher degree of representativeness of training exercises, in both SSCG and large-sided games [37].

The differences between the spatial patterns of players' roles within the same TF also underline the importance of coaches designing *supraspecific* training tasks, i.e., the specific training that goes beyond simply training with the ball [38]. *Supraspecificity* implies the design of tasks that are based not only on the football's general dynamics but also on the specific constraints of each team (e.g., coach's game model, including team formations), which has an important role in the development of the interpersonal linkages and collective synergies that underpin team performance [8].

5. Conclusions

This study exposes how *team formations* and *players' roles* influence the spatial patterns of their Voronoi cell areas. It underlines the importance of considering these features when coaches design training exercises, as they constrain players' interpersonal linkages in the establishment of team synergies and collective performance. Consequently, this study reinforces the need to train *ecologically* [39], as a pathway for players' progressive attunement to the affordances of the competitive environment, i.e., through *representative training* [11]. We believe that the assessed methods and their results can contribute to leveraging the utility of optical tracking systems in sports and ultimately to the tactical performance of high-level football teams.

Author Contributions: Conceptualisation, N.C., R.J.L. and D.A.; Data curation, N.C., R.J.L. and D.F.; Formal analysis, N.C. and R.J.L.; Funding acquisition, D.A.; Investigation, N.C., R.J.L. and D.A.; Methodology, N.C., R.J.L. and D.A.; Software, D.F.; Visualisation, N.C., R.J.L. and D.F.; Supervision, R.J.L. and D.A.; Writing—original draft, N.C.; Writing—review and editing, R.J.L. and D.A. All authors have read and agreed to the published version of the manuscript.

Funding: Duarte Araújo was partly funded by the Fundação para a Ciência e Tecnologia (grant number UIDB/00447/2020, awarded to CIPER-Centro Interdisciplinar para o Estudo da Performance Humana (unit 447). Rui J. Lopes was partly supported by the Fundação para a Ciência e Tecnologia, under Grant Number UIDB/50008/2020 attributed to Instituto de Telecomunicações.

Institutional Review Board Statement: Not applicable.

Informed Consent Statement: Not applicable.

Data Availability Statement: Third party data. Restrictions apply to the availability of these data. Data was obtained from STATS© company and are available from STATS© with their permission.

Acknowledgments: All data was kindly provided by the STATS© company.

Conflicts of Interest: The authors declare no conflict of interest.

Abbreviations

The following abbreviations are used in this manuscript:

CLA	Constraints-Led Approach
EPS	Effective Play Space
GII	Game Intensity Index
KS	Kolmogrov-Smirnov
PE	Possession Episodes
RSP	Relative Space per Player
SSCG	Small-Sided and Conditioned Games
TF	Team Formation
VA	Voronoi Area
VD	Voronoi Diagrams
Note:	Players' role in 4-2-3-1 and 3-4-1-2 team formations are defined in Table 1.

References

1. Long, J.A.; Nelson, T.A. A review of quantitative methods for movement data. *Int. J. Geogr. Inf. Sci.* **2013**, *27*, 292–318. [CrossRef]
2. Miller, H.J.; Dodge, S.; Miller, J.; Bohrer, G. Towards an integrated science of movement: Converging research on animal movement ecology and human mobility science. *Int. J. Geogr. Inf. Sci.* **2019**, *33*, 855–876. [CrossRef] [PubMed]
3. Rico-González, M.; Los Arcos, A.; Nakamura, F.Y.; Moura, F.A.; Pino-Ortega, J. The use of technology and sampling frequency to measure variables of tactical positioning in team sports: A systematic review. *Int. J.* **2020**, *28*, 279–292. [CrossRef] [PubMed]
4. Memmert, D.; Lemmink, K.A.; Sampaio, J. Current Approaches to Tactical Performance Analyses in Soccer Using Position Data. *Sports Med.* **2017**, *47*, 1–10. [CrossRef] [PubMed]
5. Memmert, D.; Rein, R. Match analysis, big data and tactics: Current trends in elite soccer. *Dtsch. Z. Fur Sportmed.* **2018**, *69*, 65–72. [CrossRef]
6. Hewitt, A.; Greenham, G.; Norton, K. Game style in soccer: What is it and can we quantify it? *Int. J. Perform. Anal. Sport* **2016**, *16*, 355–372. [CrossRef]
7. Araújo, D.; Silva, P.; Ramos, J. Affordance-based decisions guide team synergies during match performance. *Res. Phys. Educ. Sport Health* **2014**, *3*, 19–26.
8. Araújo, D.; Davids, K. Team synergies in sport: Theory and measures. *Front. Psychol.* **2016**, *7*, 1–13. [CrossRef] [PubMed]
9. Fonseca, S.; Milho, J.; Travassos, B.; Araújo, D. Spatial dynamics of team sports exposed by Voronoi diagrams. *Hum. Mov. Sci.* **2012**, *31*, 1652–1659. [CrossRef]
10. Silva, P.; Vilar, L.; Davids, K.; Araújo, D.; Garganta, J. Sports teams as complex adaptive systems: Manipulating player numbers shapes behaviours during football small sided games. *Springerplus* **2016**, *5*, 191. [CrossRef]
11. Pinder, R.A.; Davids, K.; Renshaw, I.; Araújo, D. Representative learning design and functionality of research and practice in sport. *J. Sport Exerc. Psychol.* **2011**, *33*, 146–155. [CrossRef] [PubMed]
12. Renshaw, I.; Chow, J.Y. A constraint-led approach to sport and physical education pedagogy. *Phys. Educ. Sport Pedagog.* **2019**, *24*, 103–116. [CrossRef]
13. Ric, A.; Torrents, C.; Gonçalves, B.; Torres-Ronda, L.; Sampaio, J.; Hristovski, R. Dynamics of tactical behaviour in association football when manipulating players' space of interaction. *PLoS ONE* **2017**, *12*, e0180773. [CrossRef] [PubMed]
14. Chow, J.; Davids, K.; Renshaw, I.; Button, C. The Acquisition of Movement Skill in Children through Nonlinear Pedagogy. In *Conditions of Children's Talent Development in Sport*; FIT: Morgantown, WV, USA, 2013; pp. 41–59.
15. Fradua, L.; Zubillaga, A.; Caro, Ó.; Iván Fernández-García, Á.; Ruiz-Ruiz, C.; Tenga, A. Designing small-sided games for training tactical aspects in soccer: Extrapolating pitch sizes from full-size professional matches. *J. Sport Sci.* **2013**, *31*, 573–581. [CrossRef] [PubMed]
16. Silva, P.; Esteves, P.; Correia, V.; Davids, K.; Araújo, D.; Garganta, J. Effects of manipulations of player numbers vs. Field dimensions on inter-individual coordination during small-sided games in youth football. *Int. J. Perform. Anal. Sport* **2015**, *15*, 641–659. [CrossRef]
17. Silva, P.; Travassos, B.; Vilar, L.; Aguiar, P.; Davids, K.; Araújo, D.; Garganta, J. Numerical relations and skill level constrain co-adaptive behaviors of agents in sports teams. *PLoS ONE* **2014**, *9*, e107112. [CrossRef]
18. Carrilho, D.; Couceiro, M.S.; Brito, J.; Figueiredo, P.; Lopes, R.J.; Araújo, D. Using optical tracking system data to measure team synergic behavior: Synchronization of player-ball-goal angles in a football match. *Sensors* **2020**, *20*, 4990. [CrossRef]
19. McLean, S.; Salmon, P.M.; Gorman, A.D.; Wickham, J.; Berber, E.; Solomon, C. The effect of playing formation on the passing network characteristics of a professional football team. *Hum. Mov.* **2018**, *2018*, 14–22. [CrossRef]
20. Müller-Budack, E.; Theiner, J.; Rein, R.; Ewerth, R. "Does 4-4-2 exist?"—An analytics approach to understand and classify football team formations in single match situations. In Proceedings of the MMSports 2019—Proceedings of the 2nd International Workshop on Multimedia Content Analysis in Sports, Co-Located with MM 2019, Nice, France, 25 October 2019; pp. 25–33. [CrossRef]
21. Riezebos, M. Identifying Play Styles of Football Players Based on Match Event Data. Ph.D. Thesis, Erasmus School of Economics, Erasmus University Rotterdam, Rotterdam, The Netherlands, 2021.
22. Memmert, D.; Raabe, D.; Schwab, S.; Rein, R. A tactical comparison of the 4−2−3−1 and 3−5−2 formation in soccer: A theory-oriented, experimental approach based on positional data in an 11 vs. 11 game set-up. *PLoS ONE* **2019**, *14*, e0210191. [CrossRef]
23. Merlin, M.; Cunha, S.A.; Moura, F.A.; Torres, R.d.S.; Gonçalves, B.; Sampaio, J. Exploring the determinants of success in different clusters of ball possession sequences in soccer. *Res. Sport Med.* **2020**, *28*, 339–350. [CrossRef]
24. Fernández, J.; Bornn, L.; Cervone, D. Decomposing the Immeasurable Sport: A deep learning expected possession value framework for soccer. In Proceedings of the MIT Sloan Sports Analytics Conference, Boston, MA, USA, 1–2 March 2019; pp. 1–18.
25. Yue, Z.; Broich, H.; Seifriz, F.; Mester, J. Mathematical analysis of a soccer game. Part I: Individual and collective behaviors. *Stud. Appl. Math.* **2008**, *121*, 223–243. [CrossRef]
26. Efthimiou, C.J. The Voronoi Diagram in Soccer: A theoretical study to measure dominance space. *arXiv* **2021**, arXiv:2107.05714.
27. Di Salvo, V.; Collins, A.; McNeill, B.; Cardinale, M. Validation of Prozone®: A new video-based performance analysis system. *Int. J. Perform. Anal. Sport* **2006**, *6*, 108–119. [CrossRef]
28. Duarte, R.; Araújo, D.; Folgado, H.; Esteves, P.; Marques, P.; Davids, K. Capturing complex, non-linear team behaviours during competitive football performance. *J. Syst. Sci. Complex.* **2013**, *26*, 62–72. [CrossRef]

29. Carling, C.; Dupont, G. Are declines in physical performance associated with a reduction in skill-related performance during professional soccer match-play? *J. Sports Sci.* **2011**, *29*, 63–71. [CrossRef]
30. Bradley, P.S.; Carling, C.; Archer, D.; Roberts, J.; Dodds, A.; di Mascio, M.; Paul, D.; Diaz, A.G.; Peart, D.; Krustrup, P. The effect of playing formation on high-intensity running and technical profiles in English FA premier League soccer matches. *J. Sports Sci.* **2011**, *29*, 821–830. [CrossRef]
31. *STATS Playing Styles, Definitions Manual*; STATS, Newry, Co. Down: Newry, UK, 2019. Available online: https://www.statsperform.com/resource/stats-playing-styles-introduction/ (accessed on 1 November 2022).
32. Kim, S. Voronoi Analysis of a Soccer Game. *Nonlinear Anal. Model. Control* **2004**, *9*, 233–240. [CrossRef]
33. Santana, J. Representação da área de Responsabilidade de Jogadores de Futebol através do Diagrama de Voronoi. Ph.D. Thesis, Universidade Estadual de Campinas, Campinas, Brazil, 2011.
34. Tallapragada, P.; Sudarsanam, S. A globally stable attractor that is locally unstable everywhere. *AIP Adv.* **2017**, *7*, 125012. [CrossRef]
35. Araujo, D.; Davids, K.; Chow, J.; Passos, P. The development of decision making skill in sport: An ecological dynamics perspective. In *Perspectives on Cognition and Action in Sport*; Nova Science Publishers, Inc.: New York, NY, USA, 2009; pp. 2–12.
36. Olthof, S.B.; Frencken, W.G.; Lemmink, K.A. Match-derived relative pitch area changes the physical and team tactical performance of elite soccer players in small-sided soccer games. *J. Sport Sci.* **2018**, *36*, 1557–1563. [CrossRef]
37. Gonçalves, B.; Esteves, P.; Folgado, H.; Ric, A.; Torrents, C.; Sampaio, J. Effects of Pitch Area-Restrictions on Tactical Behavior, Physical, and Physiological Performances in Soccer Large-Sided Games. *J. Strength Cond. Res.* **2017**, *31*, 2398–2408. [CrossRef]
38. Tamarit, X. *Periodización Táctica vs. Periodización Táctica*; MB Football: Valência, Spain, 2013.
39. Davids, K.; Araújo, D.; Shuttleworth, R. Applications of Dynamical Systems Theory to Football. In *Science and Football V: The Proceedings of the Fifth World Congress on Science and Football*; Routledge: London, UK, 2005; pp. 537–550.

MDPI AG
Grosspeteranlage 5
4052 Basel
Switzerland
Tel.: +41 61 683 77 34

Sensors Editorial Office
E-mail: sensors@mdpi.com
www.mdpi.com/journal/sensors

www.ingramcontent.com/pod-product-compliance
Lightning Source LLC
Chambersburg PA
CBHW041342210926
51850CB00013B/2960